SCIENCE

AND

THE HUMAN MIND

A CRITICAL AND HISTORICAL ACCOUNT OF THE
DEVELOPMENT OF NATURAL KNOWLEDGE

SIR WILLIAM CECIL DAMPIER

GIFT *Certificate*

TO:

_____

_____

FROM:

_____

_____

DATE: _____

Would you like to buy a copy of
SCIENCE AND THE HUMAN MIND ?

Please visit:
http://www.diamondbooks.ca/books

SCIENCE

AND
THE HUMAN MIND

A CRITICAL AND HISTORICAL ACCOUNT OF THE
DEVELOPMENT OF NATURAL KNOWLEDGE

BY
SIR WILLIAM CECIL DAMPIER

(*Formerly,* WHETHAM)

Sc.D., F.R.S.

(December 27, 1867 – December 11, 1952)

Sir William Cecil Dampier FRS   was a British scientist, agriculturist, and science historian who developed a method of extracting lactose (milk sugar) from whey.

DIAMOND™
BOOKS

www.diamondbooks.ca

TORONTO, CANADA – 2017

DIAMOND **BOOKS** - CANADA

BOOKS

Toronto, ON, CANADA
http://www.**diamondbooks**.ca

BIBLIOGRAPHIC INFORMATION

' SCIENCE AND THE HUMAN MIND ' was first
published in 1912, by **LONGMANS, GREEN, & CO. LTD.**
LONDON.

PUBLISHED IN CANADA

Published in Canada by DIAMOND BOOKS - CANADA, an imprint of
DIAMOND PUBLISHERS - http://www.diamondpublishers.com

REPUBLISHED EDITION: August, 2017.

PAPERBACK EDITION : ISBN: 978-1-988942-44-5
E-BOOK EDITION : ISBN: 978-1-988942-45-2

PRINTED IN CANADA

SCIENCE AND THE HUMAN MIND

A CRITICAL AND HISTORICAL ACCOUNT OF THE DEVELOPMENT OF NATURAL KNOWLEDGE

BY

WILLIAM CECIL DAMPIER WHETHAM
M.A., F.R.S.
FELLOW AND TUTOR OF TRINITY COLLEGE, CAMBRIDGE

AND

CATHERINE DURNING WHETHAM
HIS WIFE

LONGMANS, GREEN, AND CO.
39 PATERNOSTER ROW, LONDON
NEW YORK, BOMBAY, AND CALCUTTA
1912

PREFACE

THOUGH there are many histories of the different branches of science and of science itself, a general survey of the progress of natural knowledge in its relation to other fields of human thought seems not previously to have been written. This attempt to supply the need does not pretend to give a detailed account of the growth of the various sciences. It is evident that almost every section could be expanded into a volume, and each chapter heading could appropriately become the title for an exhaustive treatise. We have deliberately constrained ourselves to produce an outline, rather than the fuller study towards which we were frequently tempted. We have set out to tell in plain language the story of the separation of science from the association with theology and philosophy by which, of necessity, its origins were beset. We have tried to recount the marvellous extension of natural knowledge, following on the liberation of science ; to trace and to justify the rise of a mechanical theory of life, and to explain the recent tendency once more to recognize its limitations. Lastly, we have endeavoured to weigh the influence which, in turn, science, now admittedly supreme within its own kingdom, has had on sociology, on philosophy and on religion.

Perhaps a word of personal explanation may be forgiven. For the past ten years we have been collecting material for the subjects of the present volume. Science, like life, to be seen singly must be seen whole. To estimate the value of any piece of work, it must be fitted into the larger scheme of our knowledge of nature, and that again must find its proper place in the general record of human activity. Therefore, merely as an *apologia pro vita nostra*, it seemed inevitable, sooner or later, to write this book.

In the course of the work it has been necessary to gain a knowledge of branches of science and departments of thought beyond the confines of the somewhat restricted region of physics in which our own previous researches were carried on. At various stages, problems appeared where personal experience seemed essential. Hence, for instance, we have found ourselves at one time carrying out unsuccessful investigations on radio-activity, and, at another, putting together pedigrees showing the descent of mental or physical qualities, or breeding Mendelian hens wherewith to supply the needs of our breakfast table.

Among our former works, *The Recent Development of Physical Science* (1904), *The Family and the Nation* (1909), and *Heredity and Society* (1912) are milestones along the road of preparation ; while the *Life of*

Colonel Nathaniel Whetham (1907) was a profitable exercise in genealogical research as well as in historical method and the use of records.

Thus, whatever be the fate of this work—the result, in its present form, of the scanty leisure of ten strenuous years — we take leave of the manuscript with a sense that to us it sums up a period of widening mental outlook and of increasing interest in the varied phenomena of society, of life, and of thought.

W. C. D. W.

CAMBRIDGE,   C. D. W.
September 1912.

CONTENTS

"Hark," cried the priest of old,
"Within mine ear God breathed the hidden word."
Men came and listened, whispered, shook their heads—
    "He hath not wholly heard."

    "Stay," cried the gray-haired sage,
"Within my mind the plan, laid out, I see."
His fellows drew around;—"not so," they said;
    "He has not found the key.

    "Here," cries the latest age,
"The atom breaks and life gives up her tale."
"Is the soul naught?" the world-worn spirit sighs;
    "These men must also fail."

    "Lo," wise men cry, "we stand,
Like children, picking pebbles on the shore;
God of our fathers, give us still Thy light,
    And when that fades, give more!"

DUNSTANBURGH,
    August 1912.

CHAPTER I

INTRODUCTION

The Genesis of Natural Science—The Races of Europe—
Creation and Culture—Mysticism and Dogma.

THE vast and imposing structure of modern science is one of the greatest triumphs of the human mind.

The Genesis of Natural Science. The broad yet intimate knowledge of the phenomena of Nature now available gives not only a marvellous control over natural resources, but permits an insight into the workings of natural processes far beyond the dreams of former ages.

Within the last century, physical science has brought under its sway mechanical powers which make the tools of the past five thousand years seem like the playthings of children. Vistas have been opened into the innermost recesses of the structure of matter —vistas down which the eye of man is powerless to follow the flight of his mind.

Biology has constructed consistent theories of the mechanism of living organisms. It has revealed hosts of microscopic beings, endless in number and variety, beings pregnant with good or evil for the human race ; while the study of heredity, expressed in the new science of genetics, has arisen to light the stumbling footsteps of mankind along the dark paths of social progress.

Psychology is extending every year the knowledge of mind and its processes, conscious and unconscious. Man, as a piece of work, noble in reason, infinite in faculty, is finding himself to be the foremost problem of the age. New light has been thrown on the theory of knowledge and on the difficult problem of how knowledge at all comes to be possible to a being whose frame is but the quintessence of dust. For the human mind is the agent which has built up the goodly edifice of applied thought to be examined in the succeeding pages, and rightly claims a place in the record of the construction.

Two courses are open to us. We may take knowledge as we find it at the present hour and draw a picture of it, as it now stands ; or we may trace its growth from first beginnings in the dawn of civilization and watch the slow uprising of the various parts of the structure.

To those whose interest lies in the workings of the human mind, there is much to be learned from the development of natural science. Its inner meaning, its relations with other branches of learning, its possibilities of future development, are best revealed by some application of the historical method. Moreover, it is through a critical examination of its past history that we can obtain an insight into the value of the present stage in the evolution of any branch of the subject, and may be saved from the danger of that error, lying in wait for each generation in its turn—the error of believing that, in its own age, it has attained the final goal, at a point which is, in reality, but a step in an endless progress.

When we watch the fate of one scientific hypothesis after another as they pass across the stage from a life of active usefulness into the historical museum of intellectual curiosities, we learn that not even the youngest generation of the sons of men is to be deemed infallible. The theory by which knowledge is best advanced at any given time is not necessarily more permanent than its predecessors. The best theory is only the one which suits best the existing modes of thought, and suggests most clearly the particular direction in which advance may be made.

In natural science, as elsewhere, it is necessary to distinguish between the passing and the permanent. If the theories by which knowledge has been interpreted and extended are subject to flux and change, knowledge of nature itself stands for no transitory gain. The foundations are surely laid. Science has continued to expand, and to show interconnections and concordances between its parts which give confidence in the stability of the whole.

This steady growth in effective knowledge, and the maintenance by all competent observers of a common belief in fundamentals, are the principal features which distinguish science from philosophy and metaphysics. The attainment of general agreement is indeed the crucial point at which a subject passes from the realm of philosophic conjecture into the territory of scientific " fact."

All races at a certain stage of development treat

both science and philosophy as a branch of that much more fundamental and much more necessary mode of thought—religion. Philosophy indeed is but the attempt on one side to disentangle and analyse the various elements which are involved in the concepts of religion. It represents an endeavour to submit the inherited traditions of a whole complex social group concerning the " nature of things," the ultimate verities, to the reasoning power of one individual member of the society, who, viewing life from his own personal standpoint, strives to find arguments in support of the current conclusions. The work of science in the field of human reason is to examine the arguments, test the evidence adduced in their support, pronounce on their validity, and to suggest new methods of attack.   Therefore it fell out in the logical course of development that the earliest astronomers were probably Chaldean priests ; while the ancient Egyptians referred the origin of all knowledge to the gods, who had revealed it to those in their service. The Babylonian legend of the creation, attached by the Hebrews to the Arabian Jehovah of Sinai, supplied a cosmogony, not only to the Jews, but also to the Christian races of Europe till quite recent years.   To the early Greeks, the sun was the flaming chariot of Phœbus ; to the Indians, the clouds were kindly cows from which milk descends as nourishing rain on the fruitful earth.

It is possible that, at an earlier stage, religion itself arose from the magic rites by which primitive men try to control, by imitating, the processes of nature, or to signalize the ceremonies of tribal initiation. Magic would then be a lineal ancestor of modern science—a pleasing line of descent, since the man of science, like the magician, seeks to control the forces of nature and recognizes that knowledge is power. As Mr Cornford says, in his book *From Religion to Philosophy*, " Science, with its practical impulse, is like magic in attempting direct control over the world, whereas religion interposes between desire and its end an uncontrollable and unknowable factor—the will of a personal God. The perpetual, if unconscious, aim of science is to avoid this circuit through the unknown, and to substitute for religious representation, involving this arbitrary factor, a closed system ruled throughout by necessity." But whatever be the outcome of the enquiries now being made by anthropologists and archæologists, it is yet too soon to attempt to trace the origins of science in the dim regions of totemism. We take up our tale when religion, still carrying, it is true, signs of more archaic modes of thought, has passed into the animistic or into the anthropomorphic form.

At the stage when we begin to recognize the germs of science, knowledge is not differentiated into definite branches. The priest is also the philosopher and the physician. The first requisite for growth is the recognition of the need for separation, and the fearless pursuit of one branch of knowledge, unhindered by the trammels of methods of thought foreign to the spirit of that particular enquiry. At a much later stage, it is possible that some of the separate streams may once more converge in a higher unity, and each help the other in pushing forward a joint flood of knowledge. But, in the beginning, clear separation is necessary. Hence the first sign of advance, whether at the dawn of Greek history or at the Renaissance of learning after the gloom of the Middle Ages, is the liberation of philosophy from religion, and the next stage is the separation of natural science from them both.

The Greek philosophers effected completely the first of these changes : they freed philosophy from dependence on theology, but they never distinguished science from philosophy.

Theories of the solar system, or of the continuous as opposed to the atomic structure of matter, were to the Greeks almost as much an affair of speculation as the nature of reality or the conception of the absolute. On a narrow basis of observation, the active mind of each Greek philosopher built an imposing superstructure of conjecture, far in advance of any possibility of demonstration.   But he had not sought these problems in any spirit of arrogance.   They were part of a legacy from the earlier stages, an inheritance of outstanding puzzles from religion and magic ; and, as time has shown, true subject matter for scientific— and therefore, at that early period, for philosophic— enquiry.   To the ancients, astronomy—except in so far as the movements of the heavenly bodies were noted and used to measure time—and physics— outside the elements of geometry, land measurement and engineering—were both in truth and fact branches of philosophy, subjects as yet beyond the range of experiment and inductive reasoning.

Yet science in its earliest days advanced far in directions where immediate practical application supplied the necessary stimulus. Geometry, established on a basis of experimental knowledge, won by the need of surveying and measuring land, developed on its true method of logical deduction from a few self-evident axioms. A beginning was also made in the co-ordination of information gleaned by experience of animals and plants, and in the arts of agriculture and medicine. But, in spite of these and other successful applications, there was no conception of the power that lay in the pursuit of learning for its own sake, in the possession of organized knowledge ; and natural science remained a strange mixture of the obvious and the impossible, in which the lore of the priest, the ingenuity of the craftsman, and the fevered dreams of the magician were inextricably blended with the self-centred reason of the philosopher. Not until, following the decay of the Roman Empire, the Dark Ages had passed into the twilight of Mediævalism, did the attitude of mind begin to show signs of change ; and the dawn of the modern world was breaking in the era of the Renaissance before natural science took its stand on the firm ground of slowly won observation. Then, ceasing to be speculative philosophy, tossed about by every wind of doctrine, it became an independent and progressive branch of knowledge, developed by the healthy interaction of inductive observation and deductive reasoning.

This chronicle sums up the growth of science and explains the slow retreat of philosophy, of which the claims once filled the schools of Greece and Western Europe. While a subject is not within reach of that union of observation with logical or mathematical deduction which constitutes science, as we know it now, while it is only amenable to speculative treatment and imaginative analysis, the philosophers retain their jurisdiction. When, based on fact and supported by experiment, it passes over dry-shod to science, the philosopher's work is done ; and, if he be wise, he turns to other problems. The realm of metaphysics is always contracting ; but each successive concentration gives more power of intensive attack on the deeper and better defined problems which remain behind. The philosopher is continually losing to science ground which he has surveyed for himself, and is always gaining by his loss.

The debt which science owes to philosophy is twofold. From philosophy it learns its limitations and its interrelations ; from philosophy it takes over one problem after another, often in an advanced state of preparation for mathematical or experimental treatment. Philosophy may recognize the existence of the problem and formulate the possible solutions. Science alone can decide between them. Greek thinkers could point out that matter must be either continuous or atomic, and follow some way along the path of logical deduction from each hypothesis. But each philosopher was at liberty to take either alternative, and was powerless to convince his neighbour of the justness of his cause. It needed the slow accumulation of the definite experimental knowledge of chemical combination, available at the opening of the nineteenth century, to enable Dalton and Avogadro to bring forward overwhelming evidence in favour of the atomic theory, and to establish it in the name of science. It needed the marvellous advance of another hundred years to demonstrate directly to human eyes the existence and movement of the individual atom.

We must now endeavour to place on record some aspects of the human mind, in its interrelation to The Races of the growth of science. Let us admit, in Europe. the first place, that it is possible for a race to attain a high standard of civilization without assigning any appreciable part in its development to the effects of organized scientific discovery, which is indeed mainly a creation of the last three or four centuries. The knowledge of the workings of nature and the interest in natural phenomena have fluctuated from age to age, and from nation to nation. Some races of high standing seem to be entirely deficient in the necessary faculties either of observation or experiment. From considerations such as these, we must probably believe the growth of natural science to be dependent ultimately on some peculiarity of brain convolution and quality of mind inherent in certain individuals or types of the human species, and capable of stimulation by appropriate circumstances. The properties of the human mind, in this as in everything else, are but an indication of some biological factor.

When we pursue the problem a little further, it seems possible fairly to treat European science as a separate creation of persons or races dwelling in or derived ancestrally from the one continent. Natural science as we know it now is almost entirely the product of the races of Western and North-Western Europe, the part where the dominant elements of the nations belong to a distinct, well-defined stock. The science of ancient Greece was probably less isolated, and certainly received considerable accretions, for good or evil, from the civilizations of Assyria and Egypt. But, whether we are considering ancient or modern European

science, it is permissible to leave out of account the
achievements of the venerable Chinese empire as
having little influence, and to dwell but briefly on the
results in mathematics and philosophy due to the
dominant nations in India. We are then concerned
chiefly with the populations of Europe, and only to
a very small extent with those of the basin of the
Eastern Mediterranean, of Assyria and Egypt, and
with the Semitic races.

And here it might be worth while to recall to mind
the unique feature in continental topography presented by the great European inland sea. The lands
that approach these waters from the northern side are
remarkable for their detached and indented character.
Peninsula and island mark the coast-line, and we are
able to attach a definite history and record of separate
achievement to such small portions of the earth's
surface as Greece, Italy, and Spain ; while the islands
of the Ægean Sea, Cyprus, Crete, Sardinia, Sicily, and
many others, have all been called upon to play a
distinct part in the pageant of civilization. Yet these
places are not so far from each other and from the surrounding land but that, to a certain extent, communication must always have been open between them.
We have therefore a condition in which isolation after
invasion or conquest has produced the biological
potentialities and eventually the definite racial
qualities which become moulded into a clear-cut type,
either of the population as a whole or, as in ancient
Greece, of the conquering and governing class. But
at the same time the people have been subject to the
intellectual stimulus introduced by contact, not
necessarily biological in the sense of cross-breeding,

but contact of mind and civilization, which produces the best effects in a race or section of a nation well prepared by unity of blood and ideals to apprehend and fertilize any new conceptions. As long as the political and economic conditions did not permit of the constant free mixing and consequent degradation of races or types, as long as ideas or the great teachers who inculcated them were the only things to circulate freely, the countries of the Mediterranean each possessed a certain intellectual distinction, developing one after the other on lines best suited to the dominant national type.

And so, at a later period, when the main stocks of the Northern races had become civilized, the coasts of northern France, of the Low Countries, Germany, and the shores of the Baltic, with the peninsulas of Denmark and Scandinavia, and the British Isles, all abutting on the navigable waters of the North Sea and English Channel, reproduced the condition which had proved so fertile in earlier ages around the Mediterranean.

In order to understand our subject, we must try to glean some idea of the differences between the inhabitants of different parts of these two areas; and though some countries will be considered briefly in turn, as we have occasion to deal with them later on, we will here consider the most general propositions.

In studying the ethnology of any continent, it is important to distinguish clearly between race, language and culture, which are not necessarily coterminous in any direction. Culture may be indigenous or acquired, the expression of an inborn need or the result of an imitative and superficial disposition. Languages have been transferred from race to race, by conquest, or by proximity and slow absorption. Of the three characteristics, race is the only one which is probably fundamental in determining the genius alike of an individual and of a nation.

When first examined, the peoples of Europe seem endless in the variety of their physical characters. Yet, by studying the shape of the head, the texture of the hair, the stature and the colouring, anthropologists have found that, as we approach three definite regions, three definite combinations of characters are found in greater and greater purity. Hence the broad outlines of European ethnology have been explained by the hypothesis of three main races, taking their respective origins in these three regions, and giving rise to the complexity of existing conditions by racial mixture and interpenetration across their lines of contact. This view is in agreement with what historical evidence is available; indeed, the history of Europe is in its essence the story of the interaction of these races with each other, and with those other peoples that touch them on the east and south.

The earliest stratum of population still represented largely in Europe were the comparatively short, dark-haired and long-headed people, seen in their greatest purity in present populations of southern and eastern Spain, southern France, Italy as far north as Rome, and to a less degree in Wales, Cornwall and West Devon, the west of Ireland and parts of Scotland.

In the north-west of Europe and in lands, continental and insular, which are washed by the North Sea, there has been, ever since historic and indeed prehistoric ages, a tall, fair-haired, long-skulled, blue- or grey-eyed people, which may be known collectively as the Germanic, Teutonic or Northern race.

It is immaterial to our present purpose to enquire whether the first-named short dark race were indeed offshoots of some original North African population, which had found its way across the Mediterranean Sea, and whether the Northern race is to be regarded ultimately as a separate development or as an immemorial modification of the Mediterranean people, created by isolation, environment, or natural or artificial selection in the colder regions of Europe.

From the earliest dawn of civilization, these two races have been coexistent, and intermingled at their lines of contact. They differ alike in bodily and mental qualities, and have made and are making history by their actions and interactions.

Between the homes of these two races comes a wedge-shaped area of upland central Europe, stretching from the Auvergne and the Cevennes in France through Switzerland and Austria towards the Balkan peninsula and the islands of the Ægean Sea, and beyond again into Asia Minor, Armenia and parts of Palestine. This region is inhabited partially by the so-called Alpine race, round-skulled and broad-headed, and intermediate in stature and colouring of hair and eyes between the other two. Men of this race, which penetrated slowly down the Rhine valley into the Low Countries, reached Great Britain, but are not found in Ireland. The special characteristic of these people is the decidedly Asiatic affinities which they display, and probably we shall not be far wrong in regarding them as the result of a slow infiltration from the East.

Such then is the framework of the population of Europe during the centuries in which we are to consider one aspect of its intellectual achievements.

A race is essentially atomic in structure ; it is made up of individual parts, which we call persons, and is Creation and interpreted by these.   But the individual Culture. life is too short to give full expression to racial possibilities.   Hence each independent and creative civilization depends for its existence and progress on that continuity of tradition and definite oneness of aim and character which are essential to produce the environment suitable for the development of the typical personalities ; a unity only to be attained through a certain purity of breeding of the effective and directing portion of the race.

Consequently, periods of chaos, of the biological intermingling of races, are seldom periods of creative power and high intellectual achievement.   A superficial culture may be attained thereby, a certain denationalized temporary civilization may ensue, but the great eras of this world's thought are records of separate, distinct nationalities, expressing themselves characteristically through great men, born and bred in appropriate circumstances.   Creation is usually the outcome of one race working out its own salvation, while culture appertains to a contact of ideals and a mixture of peoples.

Our history of scientific thought will be largely the history of individuals, and, through the individuals, of the forces, biological and social, of which they were the expression.   Hence it will be pertinent, whenever possible, to note the parentage and circumstances of birth of the men of genius whose life-work forms the subject-matter of our pages. All nations contain strains of blood, differentiated from each other either in their origin, or through long periods of segregation and inbreeding among the different sections of the people, considered geographically, socially, or industrially. But, conversely, identical racial offshoots are found permeating and linking up nations widely separated from each other ; and again, a similarity of social standing added to a common ancestral origin will produce a co-ordination of outlook in quarters which are, geographically speaking, remote from each other. Owing to the intermingling of peoples and types throughout Western Europe, and especially to the peculiar stratigraphical distribution of the Northern race as administrators and directors, it is clear that there is often greater community of intellectual interest between the corresponding sections of various similarly formed nations than between the different strata of any one country. It is due to this phenomenon that science, the creation of the Northern mind in varying environments, has always proved more cosmopolitan within its limited range than, for instance, such "popular" developments as the technical or manual arts and crafts.

The essence of the mental state favourable to the growth of science seems to be a recognition of the fact that the careful and patient study of nature is the true method to obtain a knowledge of this aspect of the Universe. "Natura enim non nisi parendo vincitur." From this attitude of mind, common to all the heroes of the romance of science, it is but a step to the religious position which regards inward spiritual experience and direct personal apprehension of God as the fundamental religious verity.

This is the position of the moderate and sane mystic, who, declaring that the Kingdom of God is within, Mysticism studies the soul in its relations with the and Dogma. rest of God's creation by an inward extension of the open-minded method of experience proper to natural science. The passionate love of nature, which is characteristic of such great mystics as St Francis of Assisi, gives, when directed to its systematic study, an intuitive insight concerning its workings that bears other fruit in the life and work of Paracelsus, of Kepler, of Newton, and countless pioneers, who turned at times from the laborious methods of observation, experiment and mathematical calculation to fare forth in travel, " voyaging over strange seas of thought alone." " Truth and Freedom," as the birthright of each individual, were alike the watchwords of mystic and of man of science, although, setting out under the same banner, they attained thereby very different results.

If we contrast this attitude of mind with that prevalent among the peoples of the south of Europe, and with that of the early inhabitants of Assyria and Egypt, we are at once struck with a complete difference of outlook. St Francis, the Northerner, like Pythagoras of Samos, man of learning and mystic, some two thousand years before him, preached to his brothers, the birds. The Southern Italian ill-treats his mules, on the understanding that they have neither souls nor feelings ; while the modern Levantine pursues the same course, dubbing the four-footed creation collectively " the unreasoning ones." The Northern race, whether in its native plains or woods, its less permanent resting-places in the hill cities of the Apennines and the plain of Lombardy, on the shores of Asia Minor and Magna Græcia, or in the Thracian wilds and amid the jealous rivalries of the Hellenic states, preserves its racial characteristics. The Northern mind is not a thing apart from nature, but readily acknowledges his kinships, bestowing spirits and souls of like kind to his own on the animate and inanimate objects by which he is surrounded.

" Images and temples and altars," writes Herodotus when describing the ancient Persian religion, the creation of another branch of the Northern race, " it is not in their law to set up—nay, they count them fools who make such, because they do not hold the gods to be man-shaped, as the Greeks do. Their habit is to sacrifice to Zeus, going up to the tops of the highest mountains, holding all the round of the sky to be Zeus. They sacrifice to the sun, moon, earth, fire, water, and the winds."

The Southerner in his pagan moods reverses the process, and to express his religious convictions adds bestial countenances to the human form in order to obtain his inhuman gods. Such differences as these represent unfathomable chasms of racial antipathy.

Throughout the ages, the Southern races appear to have emphasized the necessity of form, of definite and concrete statement such as finds its highest expression in clear-cut dogma. There is a finite tree of knowledge in their philosophy, attainable somewhere, yet impious for the ordinary mortal to aspire to. Religion becomes an affair of the priesthood, of those

2

set aside and devoted to the service, the propitiation, or, it may well be, the outwitting of the gods.   There is a desire to get the conflict over, ascertain the limitation of human will and power, produce a clear-cut scheme of salvation, set the machinery in order under suitable supervision, so that humanity may continue with its worldly avocations and cease from troubling about his internal relations with God and the Universe.

The Jewish Law, the Roman Empire, the Ultramontane Roman Catholic Church, represent the culminating achievements of the advocates of universalist form and dogma.   In none of these instances can criticism be tolerated, nor is conscious expansion of thought permissible.   It is symbolic of the two opposite attitudes of mind, that St Peter, the Jew and upholder of the Law, with his very concrete keys of the Babylonian heaven and hell, should rule in Rome, the religious gathering-point of the Southern race, while St Paul, with his Hellenic affinities and mystical outlook, should hold sway in the cathedral church of the Northern metropolis.

Now the thought, the essence of which is to create, cannot long brook rigid forms or the constraint of unchanging dogma, and the soul which acknowledges a relationship with Nature demands the liberty to consort freely with its kith and kin.   Hence it has been said that the " real high school of freedom from hieratic and historical shackles is mysticism, the *philosophica teutonica*, as it was called."   And so, indeed, at one critical point in the history of thought, it will prove to have been, as we unfold our tale.   The light by which science ultimately advanced on its way has always come from the North.

But there is another mode of attack on outgrown dogma and unyielding law. If, on the one hand, authority in matters of reason has been undermined by a natural mysticism, on the other it has been destroyed by a rational scepticism, which was the weapon of the Humanists of the Italian Renaissance, chiefly men of Northern descent. The Northern mind, while quick and sure in intuition, is slow and persistent in reasoning power, and pursues its age-long enemy, intellectual authority, with the unerring instinct of a bloodhound ; for the thought, the essence of which is to create and re-interpret, can make no lasting truce with the prophets and priests of the unchanging law. Erigena, Occam and Martin Luther, the great apostles of mysticism, are not more effective in the society of their day and century than Petrarch with his daring criticism, Erasmus with his *Praise of Folly* in bitter satire of the whole external paraphernalia of the Roman Church, or Voltaire and the rationalists of the eighteenth century. Once the true conception of historical events had been obtained, the free mind of the Northern race set to work to find the correct interpretation of present tendencies in the light of past history and thought.

If, therefore, in the light of recent research and of age-long experience, we are compelled to attach so much importance to the biological factors, it follows that we are bound, in the course of our present enquiry, to consider carefully the characters of the peoples among whom science grew up. Their modes of thought, their religion, their political condition, are all pertinent to the subject we have in hand. And, as we have said, since the history of science is best written in the biography of its great men, any information that can be placed on record as to their nationality or the ultimate origin of their family must not be regarded as a ministration to idle curiosity, but as a piece of our argument, playing an essential part in the structure of this book.   It is not a fortuitous chance that Origen, the earliest and most modern of the Fathers of the Church, was born at Alexandria, of Greek parents ; that Erigena was surnamed Scotus ; or that Isaac Newton, who represents the supreme triumph of mathematical and physical thought, should have been tall, fair- or ruddy-haired and grey-eyed, and should have first seen light on a Lincolnshire freehold, in the central home of that Anglo-Danish stock which has proved the most fertile nursing mother of pure science.   Whoever is unable to appreciate the inward meaning of these facts—often deemed irrelevant by folk who are prepared to see miracles everywhere—will never comprehend the interrelations of science and the human mind.

Some sciences grow from foundations set on the firm ground of the practical arts.   Others, however they may arise, take over their problems ready framed by philosophy and all possible answers formulated by different and contending schools of thought.   Some races have made natural science and the scientific frame of mind part of the birthright of their children ; and some men of science have known and described Nature with the unerring instinct of children dwelling on the perfections of a beloved parent.

Guided by such clues as these, we shall follow in detail the development of natural science in the human mind.

CHAPTER II

SCIENCE IN THE ANCIENT WORLD

Chaldea—Egypt—India—Greece—Religion and Philosophy—Early Greek
Philosophy—The Problem of Matter—The Atomists—Aristotle—
Euclid and Geometry—Archimedes and the Origins of Mechanics—
Medicine—The Failure of Rome—The Influence of Alexandria.

AT the dawn of history two civilizations stand up out of the darkness—the civilizations of Egypt and Chaldea—known to us chiefly by the remains of buildings, sculpture, and inscriptions preserved in their waste places, and by the influence of their culture and religion on the organization and beliefs of adjoining nations and races.

Chaldea.

As far back as 2500 years before Christ, standard measures of length, weight and capacity were issued under royal authority in Babylon, an indication that the Chaldeans had already grasped the importance of fixed units of physical measurement. Here we have the first known signs of that co-ordination and standardization of the knowledge of common sense and of industry which gives the surest base for the origin of science in its practical form.

The elements of arithmetic were also known to the Babylonians. The multiplication table and tables of squares and cubes, have been found among their tablets. A duodecimal system of units, making the calculation of fractions easy, existed alongside a decimal system. The circle with its subdivisions of angular measurement, the fathom, the foot and its square, the talent and the bushel—in fact, the complete system of Chaldean weights and measures—were based on the parallel use of this double system of notation.

The beginnings of geometry, too, are found in rudimentary formulæ and figures for land-surveying, once more illustrating the origin of an abstract science from the needs of everyday life. Plans of fields led to more complicated plans of towns, and even to maps of the world as then known. But actual knowledge was woven in an inextricable manner with magical conceptions. The ideas of the virtues of special numbers, their connections with the gods, and the application of geometrical diagrams to the prediction of the future, passed from Chaldea westward, and for centuries dominated European thought.

As agriculture develops among a primitive people, the importance of a knowledge of the seasons grows also. It is a remarkable fact that the great majority of cereals and cultivated food-plants originated either in China and the Far East or at the eastern end of the Mediterranean basin. As the chief part of these plants are annuals, requiring seasonal treatment and ample water-supplies, we have here probably one reason why large centres of population early grew up in the basins of the Euphrates and the Nile, in which great stress was laid on astronomical observation. The cultivation of cereals and other annual plants makes a calendar almost a necessity, and Chaldean art shows that the use of the plough was well understood. It is to Babylon that we owe the division of the year into months, days, hours, minutes and seconds, and the invention of the sundial to mark the passing hours. The Babylonian year was one of 360 days ; the necessary adjustments being made from time to time by the interposition of extra months.

The movement of the sun and planets among the fixed stars was known, the journey of the sun across the sky being mapped out into twelve divisions according to the months. Each division was named from some Babylonian god, and represented by the symbol of that deity. Thus arose the association of parts of the sky with the crab, the scorpion and other beasts, which afterwards came to be connected with the definite groups of stars which we still know by the names of the signs of the zodiac.

Astronomical observation appears in the Babylonian records more than twenty centuries before the Christian era. From these early dates the Chaldean priests had recorded on their clay tablets the aspect of the heavens night after night, the brilliancy of the atmosphere and the skill of long practice helping their natural keenness of vision. Gradually the periodicity of astronomical events was revealed to them, till the prediction of eclipses of the moon became possible.

On this basis of definite knowledge, a fantastic scheme of astrology was built up, and, indeed, regarded by the Chaldeans as the chief and most worthy object of the underlying science. Starting, doubtless, from some observed coincidences, belief in the power of predicting and even of influencing future events by observing the stars as interpreting the minds of the gods became general, and gave the Babylonian priests a power both over public and over private affairs. Each temple put together a library of astrological literature from which the methods of divination might be learnt. One such library, consisting of some seventy clay tablets, was of special repute in the seventh century before Christ, and is considered to date originally from some time about 3800 B.C. " Astronomy, as thus understood, was not merely the queen of sciences, it was the mistress of the world."

Astrology reached its zenith in Babylon about 540 B.C., and two centuries later it spread to Greece and over the known world, although in its original home it was showing signs of passing into a more rational astronomy. But Chaldean astrologers continued to be in request, while their ignorance of the elements of medicine did not prevent an almost equal faith in Chaldean sorcerers and exorcists.

These theories of magic naturally followed from the conception of the powers of nature as animate. And the special form the theories took in Chaldea arose partly from the fact that on the whole the Babylonian gods were inimical to man. Carved in the shapes of monstrous beasts, their presentments terrified the eye, while by magic alone could they be induced to delay the decrees of destiny or modify the inexorable fate which even they could not wholly avert. Dominated by such a gloomy religion, no speculative science or rational philosophy could arise.

When we turn to consider the other great civilization of early times—that of Egypt—a difference of religious attitude is seen at once to produce a different scientific spirit. While in Chaldea and, more markedly, in Assyria the gods were usually conceived as hostile to man, pursuing him in life and death with an implacable hatred, in Egypt, as in Greece, the divine powers were represented in mythology as friendly, ready to watch over, to protect and to guide mankind in life, in death and in the after-world.

Egypt.

It would be interesting to enquire what share the external conditions of their lives have in shaping the attitude of a people towards the forces of Nature and the mythology by which they endeavour to interpret the phenomena of the world and of consciousness. In Egypt, the Nile, with its regular and unfailing rise and fall, was the source of all fertility—steady, trustworthy and friendly. In Chaldea, the tempestuous and incalculable flooding of the Euphrates and the Tigris made life near their banks dangerous and uncertain. Nature was hostile, ready to sweep away man and his puny works together in one unforeseen ruin. The climate of Egypt is more equable, free from the extreme variations and violent storms that harry the highlands of Assyria and Armenia and the shores of the Persian Gulf. It is possible, too, that her isolated position both geographically and politically made Egypt less liable to the ravages of war, and to the visitations of plague and other diseases which would more easily reach Chaldea from the closely populated East.

Even with an original or partial similarity of race, these differences in natural conditions might modify profoundly the course of the development of the human mind, and, in other cases, might accentuate and intensify racial differences. In the one case, any attempt to understand or to control the elemental forces becomes an impious and useless action. Deceit and trickery by magic and sorcery, or at the best propitiation of the hostile powers by sacrificial bribes, represent the logical outcome of this view of nature. It is a good example of racial memory that compelled the Greeks to leave Prometheus, outspread on the Caucasus, punished for helping mankind with the gift of fire, and decreed that the serpent which tempted Eve should have been bred in the gardens of Asia Minor. As we see in the story of Adam, the acquirement of knowledge is a form of sacrilege ; it trenches on the divine prerogative and must be expiated by death, degradation or other punishment. But, in the second case, when the deities are friendly, any increase of man's mastery over his surroundings is approved by the tutelary powers, and is probably directed by them, since it is their good pleasure to help him on his way.

One or other of these alternative attitudes of mind predominates in every religious system, according to the race and circumstances of those who hold to it, according to the mode of its origin and the reaction of philosophic thought on its history. And, in its turn, the prevalent type of religious thought influences philosophic tendencies and scientific progress.

In Egypt, as we have said, the gods on the whole were friendly to mankind, and willing to help his efforts in the acquirement of knowledge. From the earliest times of which we have definite records,

Egyptian civilization was at a comparatively advanced stage of development. But the possibility of a long and slow growth in art and science had not dawned on the minds of men. It seemed clear that their ancestors, left to their own human resources, could never have made such discoveries as that of speech or of writing. It was therefore natural to imagine a divine intervention to account for the origin of every art, craft, or science. All advance in knowledge was thus ascribed to the revelation of the gods, and especially of Thot, represented by the Ibis or Baboon, and of his ally Mâit, the goddess of truth. Thot, one of a legendary race of divine sovereigns or legislators, was essentially a moon-god, who measured time, counted days and recorded the years. But he was also lord of the voice, master of books and inventor of writing.

This theory of the method of progress continued to be put forward officially even when it was scarcely possible that the more intelligent of the people could have been deceived by it, at all events as regards recent changes. Thus, in a text-book on medicine, a chapter on a special branch of disease, which was known to be of a later date than the rest of the work, was not referred to the learning of any physician, but was stated to have been disclosed one night to a priest who was watching by moonlight in the temple of Isis.

In arithmetic and geometry the knowledge of the Egyptians was about on a level with that of the Chaldeans. With the periodic submersion of the land beneath the waters of the Nile and the obliteration of boundaries, land measurement was of even more importance, and doubtless accounts for the origin of a science, once more referred to the benevolence of Thot.

It was Thot, too, who had invented music and drawing, and revealed their laws to mankind. It was Thot who had established a service of observers, the " watchers of the night " or the " great of sight," who followed and noted down from the temples the principal astronomical changes and sequences. But, while it may have rivalled the Chaldean astronomy in age, the corresponding Egyptian science never reached such an advanced stage of development. Doubtless, the importance attached by the Chaldeans to astrology gave a more powerful motive for astronomical research, while the wealth and power at the disposal of a successful astrologer gave to his pure science the pecuniary resources and extended impulse which in our own times have been found greatly to aid those branches of knowledge capable of bearing practical application to the arts of life.

If Egypt lagged behind Babylon in astronomy, and possessed no astrologers with the reputation of those from Chaldea, in medicine the relative positions were reversed. Babylon possessed no school of rational medicine, but relied on sorcery and exorcism for the treatment of disease, which was referred exclusively to the action of malignant powers. But in Egypt medical knowledge was highly specialized at an early date. There were physicians trained in the priestly schools, bone-setters for the treatment of fractures, while mental diseases alone seem systematically to have been left to exorcists, who, by means of amulets and magic, drove out the evil spirits responsible for these infirmities. The art of dispensing drugs and essences had been brought to a high state of excellence, and many of the Egyptian remedies became of worldwide application. Indeed, Egyptian medicine generally had an extended influence : the Greeks acquired it, perhaps by way of Crete, and, in after ages, from Greece it passed with the rest of their learning into Western Europe.

The scientific development of the Indo-Teutonic races in Northern India early reached a very high
India.   point both in philosophy and mathematics, in which subjects their influence travelled to Greece by way of Persia and Assyria. These people, in the first stages of their settlement, possessed the open confidence in their own mission, the friendly attitude towards their divine powers, the broad sweep of thought, the restless energy, which are characteristic of their stock, whether in Asia or Europe. But, from various causes, of which, in spite of their caste or colour restrictions, intermarriage with the native races was probably the most effective, their scientific development seems to have been checked, or, at least, progressed only in certain restricted directions. The advent of Buddhism arrested their normal expansion, and turned the thoughts of their great men away from the problems presented in the ordinary course of existence. A religion took root and grew up among them, which, emphasizing the transitoriness and vanity of earthly existence, made self-annihilation and loss of individuality a condition in the attainment of spiritual completion. This attitude of mind, apparently a concentrated and overbalanced form of mysticism, by distracting the attention from all immediate surroundings, must at once arrest that desire for material improvement which is often the

mainspring of an advance in scientific knowledge. It is a mental state as extreme in the one direction as the frankly materialistic view of life is in the opposite way.   Each fails ultimately in that it takes no account of the dual nature of mankind, that it ignores the interconnection of the mind and the body, of the individual and the society.

But there is no doubt that at all times the philosophy of India indirectly influenced the schools of thought of Asia Minor; and it is certain that, during the Arab domination in the lands of the Eastern Mediterranean, the mathematics and medicine of India mingled with the stores saved from Greece and Rome, and re-entered the schools of Western Europe by way of Spain and Constantinople.

On Greece all the separate streams of knowledge of the ancient world converged, there to be filtered and purified, and turned into new and more fruitful channels by the marvellous genius of the first European race to emerge from obscurity.

Greece.

To understand the genesis of the natural philosophy of the Greeks, a natural philosophy which formulated so many of the problems afterwards attacked by science, and suggested tentatively so many successful solutions, we must consider briefly the Greek populations, their religion, and the physical and social conditions of their existence.

It is probable that the effective part of the early inhabitants of Greece were a tribe of conquerors belonging to the great Northern race whose chief home seems to have been on the shores of the Baltic.   This dominant race, descending through Central Europe, was superimposed in Greece on earlier inhabitants among whom men of the Mediterranean stock predominated ; while in Asia Minor, and some of the islands of the Ægean Sea, it encountered chiefly Alpine or Armenoid peoples, mingling with the darker folk only on the coast-line. The so-called Minoan civilization, of which the centre point seems to have been discovered in the excavations of Cnossus in Crete, may have represented the earliest achievements of the Mediterranean peoples. It seems to have been influenced considerably by Egyptian culture ; while the contemporary and succeeding bronze age, with a civilization which seems to start at an advanced stage of development in some of the cities of the mainland and the islands, has been associated with men of the round-skulled type and may represent an early descent of the Alpine stock, driven forward by the Northern race, who were shortly to follow on their footsteps.

As we have said, the Mediterranean folk were and are dark-haired and long-headed, while their Northern conquerors were fair-haired and long-skulled. The Mediterraneans buried their dead, while the heroes of Homer, like the warriors of the North, passed to the other world amid the flames of a funeral pyre. If this hypothesis be correct, the peoples of ancient Greece consisted of much the same elements, but in different proportions, as those of England when the earlier dark race had been overlaid by the influx of Saxon and Dane and Northman ; and a natural affinity of thought and outlook might be anticipated therefrom. Moreover, the stimulus of an expanding colonial dominion is common to both. The chiefs

of the Homeric age had settlements on all the lands
which fringe the Mediterranean Sea, and Southern
Italy became known as Magna Græcia. Intercourse
with these oversea outposts, and with the divergent
nations that surrounded them, doubtless played a
great part in developing the economic wealth of the
nation and the innate powers of the Greek mind.

While the citizens of the Greek states, the warriors
and the traders, were chiefly or exclusively of the
Northern stock, the Mediterranean people, retaining
their ancient culture and primitive conceptions, still
tilled the soil and appeared as the chief element of
that slave population on which Greek civilization was
built up.

In Greek religion we seem to catch traces of this
dual origin. There is an underworld of spirits of
doubtful or hostile intention towards men, and underlying this again are vestiges of totemistic beliefs, and
of that system of magic which springs naturally from
the confusion between the life of nature and the life
of the tribe, and seems prior to any idea of gods.
Here, probably, we have the influence of the religious
attitude which characterized the earlier dispossessed
population.   But, in the Homeric poems, which sing
the heroes of the conquering Northern race, we find a
new joyousness of outlook, an attitude of friendly
relations with fully developed divine powers, who, as
the populations begin to mingle, are indeed figured
simply and naturally as supermen, and superwomen,
always interested in mankind, partisans and poli-
ticians of the most pronounced type, taking a share in
the life of the nation, its wars, its trials and its successes.
We find too the attribution of the invention of the arts

and sciences to the gods and demi-gods, who are always ready to appear among men, to build their cities, and to beget heroes to be the fathers of the nations, and to outwit the ancient shadowy powers, looming distressfully in the background.

As far back as the year 546 before Christ, the philosophical poet, Xenophanes of Colophon, recognized that, whether or no it be true that God made man in His own image, it is quite certain that man makes gods in his. And from the gods of the old Greek mythology we get an insight into the genius of the Greek that nothing else can give. We see the picture of a race, false, boastful and licentious perhaps, but with a sense of beauty, a confident joy in life, a warmth of affection that bespeak a gallant, vigorous, open-hearted, conquering people, a people of extraordinarily brilliant original intellectual endowment, tempered and purified by the rigours of the North, and then placed in a land of glorious beauty, where the wine-dark sea brought the trade and knowledge of the world to their doors, where the climate smiled upon their fortified homesteads, where abundant slaves made life easy, and gave leisure for the growth of the highest forms of philosophy, literature and art.

The main function of the Greek religion, as of many others, was to interpret nature and its processes Religion and in terms which could be understood— Philosophy. to make man feel at home in the world. The animistic conceptions in which it was expressed were of unusual beauty and insight. Each fountain lived in its nymph, each wood in its dryad.

3

The grain-bearing earth was personified as Demeter ; the unharvested sea came to life in Poseidon the earth-shaker.

From generation to generation the divine figures are multiplied and more clearly delineated, new attributes are assigned to them, cycles of stories cluster round each name. We watch a continual process of evolution. Each poet is free to adapt the myths to his own purpose ; to introduce a recovered legend, to weave a new allegory ; to re-interpret at his will. As the ages pass, and the intellect masters the senses, a need for a higher creed is felt, until at length Æschylus and Sophocles evolve out of the older crude polytheism the idea of a single, supreme and righteous Zeus. All this is wrought quite naturally, with hardly an idea of innovation, by those whose object is to preserve, purify and expound the old faiths. The process culminates in the metaphysical reconstruction of religion by the genius of Plato, which, founded on the social and ethical needs of the Hellenic people in the day of their greatest triumph, rises to those highest flights of mysticism which afterwards so profoundly influenced the theology of Christianity.

But, alongside this process of conservative evolution, a simultaneous sceptical criticism was going on. A religion so frankly anthropomorphic appealed rather to the imagination than to the intellect, and its weakness on the philosophical side became apparent when growing doubt ventured to express itself more openly. This very weakness, coupled with the essential freedom of intellectual outlook of the Greek world, led to a natural and metaphysical philosophy, which, even from the earliest times, was untrammelled by all theological preconceptions.

The beginnings of mediæval philosophy and science had to work under the hampering condition of a religious system of cast-iron dogma, which completely dominated the thoughts of all men, even those working at other problems, and supplied to all physical and biological questions, as well as to those of metaphysics and theology, an interpretation not to be gainsaid. Hence, in the Middle Ages and at the Renaissance, philosophy and science had to struggle for freedom almost before they could begin their struggle for existence.

In the growth of Greek natural philosophy the circumstances were different. All things are comparative. Outward obstacles, of course, were not wanting. Anaxagoras was driven from Athens as an atheist, and the same charge was one of the counts in the indictment of Socrates. Aristophanes could not refrain his inimitable jesting at the expense of the physical speculations current in his day. If as nothing compared with the universal and vigilant outward oppression and inward scrutiny of the Roman Church, and later of its Inquisition, these obstacles doubtless had a very real effect.

But, in the Middle Ages, a barrier to the birth and development of rational ideas of nature greater than the persuasions of the rack and the stake must be recognized in the inevitable mental attitude even of the physical enquirers themselves. The mediæval mind was completely dominated by the ideas of a theology formulated in the last classical ages, and crystallized into greater definiteness and rigidity by the acute dialectic of mediæval scholasticism.   Save in the rarest cases, the mediæval mind could not shake off the trammels which compassed it about in every direction. The tyranny of Rome and its dogmatic school pressed more hardly inwardly over the workings of the mind than outwardly over the expression of its opinions.

This inward intellectual constraint, this permanent bent of the mind, in deference to traditional bonds which could not be cast off and could scarcely be loosened, seems to have been much less powerful among the Greeks.   The fluidity of their religion, the variety of its ever-changing myths, its adaptability to the needs of poetic and artistic beauty, its readiness to incorporate and to adorn new ideas, led to a freedom and openness of intellectual outlook quite foreign to the Middle Ages.

The natural bias of the two religions—the Greek, the expression of the genius of the Northern race set free in a Southern atmosphere ; the Mediæval, the warping and distorting of the divine truths of Christianity through the incorporation of late Jewish dogma by the bewildered minds of the denizens of the early centuries of chaos—this natural bias was exaggerated by the difference in social conditions.   The full and fairly secure life of a Greek city contrasts with the painful emergence of some sort of safety and ordered though narrow existence under the protection of the Church from the welter of economic ruin and social confusion of the Dark Ages.   It is no wonder that in one case we get freedom and breadth of outlook, in the other narrow preconceived ideas and rigid dogmatism to which the mind clings for safety and dares not contemplate an escape.

But, if the fluidity of Greek religion gave one advantage to philosophic thought, the very weakness of that religion on the rational side gave another even greater support. None but the most powerful and acute of mediæval minds could pierce the encircling gloom of mediæval Christianity, and detect the intellectual failure of its philosophic and scientific system. But any Greek philosopher at once could see that the beautiful nature-myths of his national faith were useless as a basis for physical speculation on the origin and nature of the world. Hence, even from the first, we find philosophers building on a frankly rationalist theory. They differ in methods, in results, but all alike gently lay aside the legendary origins, reject the supernatural, and assume as an axiom for the world of thought some physical causation.

Now, whatever be the truth about metaphysical reality, natural science can only advance by working from hand to mouth on the assumption that each step is to be explained by rational and natural causes. Hence the success, such as it was, of the Greek philosophers.

When the Greek states developed and outgrew their earlier days, the geographical position of the country, as well as its economic needs, brought them into contact with the two older civilizations that we have already briefly considered. Greece seems to have drawn much of its philosophy, its mathematics and its astronomy from Asiatic sources, while its medicine and its geometry came from Egypt, possibly by way of Crete. The area of Hellenic thought, the product of this mingling of ideas, moved gradually westwards, beginning in the Ionian islands of the Ægean Sea, where the Hellenes first met the men of Asiatic descent. It then reached the zenith of its philosophical and artistic development at Athens and in the cities of the mainland, and spread gradually to the colonies of greater Greece in Sicily and South Italy, where the practical genius of Archimedes marks its highest achievement in natural knowledge.

The first known European school of thought to break away definitely from the mythological tradition
Early Greek
Philosophy.
was that of the Ionian philosophers. The position of the cities of Ionia and the adjacent isles as the starting-point of Hellenic philosophy is of great interest. Either their situation at an extremity of the region overrun by the northern invaders enabled them at an early period to .recover the settled course of steady development, free for a while from further intrusion, or else the character of the existing population was already extraordinarily favourable to the development of philosophic thought. Possibly we may here encounter the effects of some earlier unrecorded migration of a similar racial stock which had already prepared the soil and leavened the inhabitants. Thales of Miletus (580 B.C.) is the earliest of the Ionian teachers whose fame has reached us. He taught that water, or moisture, was the essence of all things, that everything possesses a principle of activity—has " soul in it "—and that this " soul " is divine in origin, or, at any rate, is superhuman. The importance of the Milesian school of philosophy lies in the fact that it possessed a unity of purpose and definite aim, setting out, by means of such research

as was then possible, to answer the question, " Of what and in what way is the world made ? "

Thales was followed by Anaximenes, who seems to have been the first clearly to recognize that the heavens revolved round the pole star, and to draw the conclusion that the visible dome of the sky was half of a complete sphere at the centre of which was the earth. Till Anaximenes gained this new outlook, the earth had been imagined as a floor with a solid base of limitless depth.   It was now represented as a flattened cylinder, floating within the celestial sphere, which, carrying the stars with it as fixed luminous points, revolved about the earth as the centre of all things.   Thus, in its day, the now discredited geocentric theory was an immense advance over the mythological ideas which preceded it.   A false hypothesis, if it serve as a guide for further enquiry, may, at the right stage of science, be as useful as, or more useful than, a truer one for which acceptable evidence is not yet at hand.

In this case, though a better theory was shortly forthcoming, the time was not ripe, and, for many centuries, mankind returned to the earlier view as giving a more convincing picture.   The school of Pythagoras (c. 530) has been held to favour a mystical attitude of mind, as opposed to the rationalistic tendencies of the men of Miletus.   A religious reformation was probably involved in their teachings, and in them we first meet with the conception of contrasting principles, Good and Evil, Light and Darkness.   The Pythagoreans replaced Anaximenes' moving sky by a moving earth, which was imagined to revolve round a central point fixed in space like a stone at the end of a string, presenting its inhabited outer face successively to each part of the surrounding sky.

In the fourth century before Christ, geographical discovery was advanced greatly, especially by the voyages of Pythias round Britain towards the north polar seas. The knowledge thus gained of the variation with latitude of day and night, etc., led to the simpler conception of the revolution of the earth on its own axis, while Aristarchus of Samos (280–250 B.C.) held that the sun was larger than the earth, formed the true centre of our system, and was one among the countless fixed stars of heaven. But this heliocentric theory was too much in advance of the time to carry general assent. The mass of mankind, even the majority of philosophers, still considered the centre of the Universe to be the earth, whether as a floating ball round which the heavens revolved, or even as the fixed, stable, bottomless solid it seemed to the senses. The centuries from the days of Aristarchus to those of Copernicus had to pass before enough evidence accumulated to force men to revive the speculative view of Aristarchus, and to establish it in the light of all the new evidence as the universally accepted theory of science.

The authority of Aristotle was too great for the new theory of Aristarchus, and, about 130 B.C. Hipparchus, the inventor of trigonometry, developed the geocentric theory into a form which, expounded by Ptolemy of Alexandria (*fl.* 127–151 A.D.), held the field till the sixteenth century of our era.

The theory of Hipparchus, though erroneous in its underlying assumption and in its results, was founded on sound methods of induction. Accepting the earth as the centre, Hipparchus showed that the apparent motion of the sun could be explained by supposing that it was carried round a central point in an orbit or epicycle, while this orbit, sun and all, was carried round the earth in an immensely larger circular orbit or cycle. From direct observation, solar tables could then be calculated, from which the sun's position at any future time could be predicted, and even solar and lunar eclipses foretold with some accuracy. These ideas were extended with success to the moon also, Babylonian records being pressed into service for the dates of past eclipses.

According to the accepted view, continued motion needed a continued moving force, and thus it was necessary to suppose the skies filled with crystal circles, which, cycles and epicycles, carried round the heavenly bodies as they revolved.

It is easy to ridicule this scheme of astronomy in the light of modern times ; yet the fact remains that, false as the theory was, it served successfully to interpret knowledge for many centuries, and guided the labours of such a competent astronomer as Ptolemy. The theory most fruitful at a given time is not always that which eventually survives, but may be too much in advance of the age to be usefully employed.

If astronomical phenomena are the more striking, and therefore the first to arrest attention, the problem The Problem of the nature of matter cries equally for of Matter. an explanation to thoughtful minds. And so we find the Ionian philosophers tracing the changes of substance from earth and water to the bodies of plants and animals, and back again to earth and water. They began to realize the conception of the indestructibility of matter, and speculated on the possibility of a single " element," a common basis of all substances. But the rival hypothesis of the Sicilian philosopher Empedocles (450 B.C.), the wonder-worker, appealed more to contemporary thought. Empedocles held that the primary elements were earth, water, air and fire—a solid, a liquid, a gas, and a type of matter still rarer than the gaseous. These four elements were combined throughout the Universe in different pro-portions under the influence of two contrasted divine powers, one attractive and one repulsive, which the ordinary eye sees working among men as love and hatred, a return to the conceptions of Pythagoras. Empedocles illustrates in a most forcible manner the essential unity of the mystic and the man of science, on which we have already dwelt. He was able to distinguish between the body substances of his four corporeal " elements," and the soul substance of his two divine forces. He separated energy from matter, and, in so doing, took a real step in advance. This doctrine of four elements held sway over the minds of men till modern chemistry was brought up against the eighty or more different types of irresolvable matter, and it is still found in literature now that modern physics is opening once more a glimpse into yet more fundamental depths and discovering corpuscles, common to all chemical elements—a return by the path of modern experimental methods to the lucky guess of the Ionian philosophers.

Empedocles' theory of a divine power or life passing

from element to element seems to have been a way of teaching the identity of the principles underlying every phase of development from inorganic nature to man. According to Empedocles, existence is essentially one from end to end.

Empedocles saw that, by imagining his four elements united in different proportions, he could explain all the endless kinds of different substances known to mankind.   Leucippus and Democritus developed this hypothesis into a theory of atoms which we know best through the description of it given at a later date by the Latin poet Lucretius.

The Atomists.

The ground on which the atomic theory of the Greeks was founded was very different from the definite experimental facts known to Dalton and Avogadro when they formulated the atomic and molecular theories of to-day.   The modern chemists had before them exact quantitative measurements of the proportions in which chemical elements combined by weight and by volume.   These limited and definite facts led irresistibly to the idea of atoms and of molecules, and gave to them at once relative atomic and molecular weights.   The theory thus formulated was then found to conform with all the rest of the countless facts and experiences which had become the common possession of science, to be supported by other relations as successively they were discovered, and to serve as a useful guide in the study and even the prediction of new phenomena.

But the Greeks had neither the definite experimental facts to suggest the theory in the first place, nor the power of testing by experiment the consequences of the theory when deduced. The theory remained a doctrine, like the metaphysical systems of our own day, dependent on the mental attitude of its originator and his followers, and liable to be upset and replaced from the foundations by the new system of a rival philosopher. And this, indeed, is what happened.

The Greek atomists reasoned from the general knowledge of their day in the light of the prevalent metaphysical ideas. When matter is divided and subdivided, do its properties remain unchanged ? Is earth always earth, and water water, however far the process of division be carried ?   In other words, are the properties of bodies ultimate facts of which no further explanation can be given ?   Or can we represent them in terms of simpler conceptions, and thus push the limits of ignorance one step further back ?

It is this attempt at a rational explanation which gives to the atomic theory of the Greeks its importance in the history of thought. According to the ideas that preceded its birth, the qualities of substances were thought to be of their essence ; the sweetness of sugar, and the colour of leaves, were as much a reality, not to be explained by reference to other facts, as the sugar and leaves themselves.

In contra-distinction to this view, which made all further enquiry useless, Democritus taught : " According to convention there is a sweet and a bitter, a hot and a cold, and according to convention there is colour. In truth there are atoms and a void."

The atoms of Democritus were uncaused and existent from eternity.   They were many in size and shape, but identical in substance. Empedocles explained differences in properties by different combinations of his four elements. Democritus went further, and referred differences in properties to differences in size, shape, position and movement of particles of the same ultimate nature. His theory, as transmitted to us by Lucretius, effects a wonderful simplification in the mental picture of nature previously held. In fact, the picture is too simple. The atomists passed unconsciously over difficulties still unsolved after the lapse of twenty-four centuries. Fearlessly they applied their theory to problems of life and consciousness which still defy explanation in mechanical terms. Confidently they believed they had left no mysteries, all blind to the great question whether the conception of atoms they had framed to describe phenomena corresponded with an ultimate reality any more nearly than did the simple sense-perceptions it was invoked to explain.

Nevertheless, the Democritan atomic theory is nearer to the views now held by physicists than any of the systems which preceded or replaced

Aristotle.

it, and its virtual suppression under the destructive criticism of Aristotle (c. 384–322) must be counted a loss to mankind.

The success of the Aristotelian philosophy shows the danger to physical theories, even though sound in themselves, when they are not founded on a broad and detailed basis of experimental fact. Because the consequences of the atomic theory did not agree with Aristotle's preconceived ideas, he rejected it altogether, and, in the absence of definite confirmatory evidence, he was able to obtain a general acceptance of his views.

As an example of Aristotle's method, his treatment of the problem of falling bodies is instructive. Democritus had taught that in a vacuum all bodies would fall at an equal rate, and that the observed differences were due to the unequal resistance of the air. This opinion is correct, though Democritus had no experimental evidence to bring forward. Aristotle accepts the statement that in a vacuum bodies would fall equally fast, but argues that such a conclusion is inconceivable, and that therefore there can never be a vacuum. With the possibility of empty space he rejects all the allied concepts of the atomic theory. If all substances were composed of the same ultimate material, Aristotle argued that they would all be heavy by nature, and nothing would be light in itself or tend to rise spontaneously. A large mass of air or fire would then be heavier than a small mass of earth or water, and the earth or water could not sink through air or fire as it is known to do.

Aristotle's error arose from the fact that, in common with other philosophers of his age, he had no notion of the conceptions now known as density or specific gravity; he failed to see that it is the weight per unit volume which is the determining factor in questions of rise and fall, and attributed the motion to an innate instinct leading everything to seek its natural resting-place. This doctrine that bodies are essentially heavy or light in themselves was accepted with the rest of Aristotle's philosophy by the schoolmen and theologians of the later Middle Ages, and his dead hand held back the flowing tide of knowledge till Galileo, in the years 1589 to 1591, appealing to actual experiments carried on from the top of the leaning tower of Pisa, showed that heavy and light bodies do fall at the same rate, and thus destroyed the Aristotelian conception of heaviness and lightness as essential qualities of bodies.

Aristotle, too, though he accepted the spherical form of the earth, maintained the geocentric theory of the Universe, and his authority did much to prevent the heliocentric theory of Aristarchus from being accepted in the next generation, and obstructed its revival by Copernicus seventeen hundred years later.

But nevertheless Aristotle was the greatest collector and systematizer of knowledge produced by the ancient world. His supreme importance in the history of science consists in the fact that, till the Renaissance of learning in modern Europe, no appreciable advance in our knowledge of nature was made in all the centuries that followed him. The task of the Dark Ages was to preserve what it could glean of his works from imperfect and incomplete abstracts; and later mediæval times spent their strength in recovering his meaning when the full text of his books reappeared in the West. Aristotle's works are an encyclopædia of the learning of his generation, and, save in physics and astronomy, he probably made real improvement in all the subjects he touched. He gives careful descriptions of animals, their life-histories and habits; he laid good foundations for the modern science of comparative anatomy, and, the first embryologist, watched the heart beating in the chick while yet in the egg. Even in physiology, though his conclusions and theories were often wrong, his methods marked a great step in advance. For instance, after giving an account of the views on respiration held by earlier naturalists, he points out that " the main reason why these writers have not given a good account of these facts is that they have no acquaintance with the internal organs, and that they did not accept the doctrine that there is a final cause for whatever Nature does. If they had asked for what purpose respiration exists in animals, and had considered this with reference to the organs, *e.g.* the gills and the lungs, they would have discovered the reason more speedily." Here the insistence on the need of observation of anatomical structure before the framing of views on the functions of organs is the point, and, in the treatment which follows, Aristotle passes in review the structure of a number of animals, and describes the action of their lungs or gills. In drawing conclusions, Aristotle had, of course, no knowledge of chemistry to help him, the idea of gases other than air was unknown, and the only change in air which could be suggested was its heating or cooling. Aristotle's theory that the object of respiration is to cool the blood by contact with air, though now seemingly absurd, was perhaps the best of which his age was capable.

Aristotle took over from Plato, his master in philosophy, many metaphysical ideas, some of which he modified in accordance with his greater knowledge of nature. Plato had no scientific insight ; his interests were philosophical. Hence perhaps arises the fact that Plato's theory of nature as a whole, and even that of his pupil Aristotle, were less in accordance with what we now hold to be the truth than the conclusions of the older nature-philosophers, though, in points of detail, Aristotle far surpassed them in knowledge.

With the more metaphysical aspects of Greek thought we are not concerned.   Yet, owing to its importance in mediæval controversy, it may be well to touch lightly on Plato's doctrine of ideas, and Aristotle's variation of it.

In nature we find numberless groups of objects more or less similar ; animal and vegetable species may stand as examples.   To explain the similarity, the mind conceives a primary type to which, in some way, the individuals have to conform.   Now, when the mind begins consciously to frame definitions and to reason about them in general terms applicable to any particular case, Plato found that the definitions and reasoning were concerned with these hypothetical types.   All natural objects are in a constant state of change ; alone the types imagined by the mind remain constant and unchangeable.   Hence Plato was led to the theory that these ideas of the mind are the only realities.   For Plato, says Aristotle, " the objects of sense were additions to the ideas, and named after them, for it was by participation in the ideas that their material namesakes existed."

To Aristotle, often immersed in the detailed study of definite individual animals or other concrete objects, this thorough-going idealism was not a convenient attitude of mind.   The influence of his master remained, but he broke away from his extreme position. While admitting the reality of the individuals, the

4

concrete objects of sense, Aristotle recognized also a secondary reality in the universals or ideas, which he held to exist in and with the objects of sense as their essence. In later times this view was developed into pure nominalism, according to which the individuals are the only realities, the universals being taken merely as names, or mental concepts. To this question we shall be brought back in dealing with mediæval thought.

Now, whatever be the truth about Plato's doctrine of ideas from a metaphysical or logical point of view—and in a modern form it is still held by many philosophers—the mental attitude which gave it birth is not adapted to further the cause of experimental science, before its exclusively logical or metaphysical import is recognized. The school of Plato, including even Aristotle, were far too prone to treat a scientific problem under a preconception of the supreme importance of words and their meanings, to consider that hot is necessarily opposed to cold, heavy to light, and to assume that bodies possessed essential qualities corresponding to one or other of these contrasted words.

The characteristic weakness of the inductive sciences among the Greeks is explicable when we examine their theories of knowledge. Aristotle, while dealing skilfully with the theory of the process of passing from particular instances of judgments of sense to general propositions, regarded this induction merely as a necessary preliminary to true science of the deductive type, which, by logical reasoning, deduces consequences from the premises reached by the former process.

The best instance of a deductive science, in fact one of the most striking results of Greek thought, is the Euclidean treatment of geometry. Euclid (*c.* 300 B.C.) collected and systematized and developed the work of preceding geometers. From a few axioms which he regarded as properties of space evident from our direct experience, he deduced by logical self-evident principles a wonderful series of propositions, in a manner which remained till quite recent years the only accepted method. As the science of the space known to our senses, its results marked a permanent step in advance, which, unlike some other products of Greek thought, had never to be retraced.

Euclid and Geometry.

The foundations of the sciences of mechanics and hydrostatics are to be sought in the work of Archimedes of Syracuse (287–212 B.C.), whose work, more than that of any other Greek, shows the true modern spirit of experimental enquiry in which hypotheses are set forth only to be tested by experiment. The idea of the relative density of bodies, which, as we have seen, was unknown to Aristotle, was first formulated by Archimedes, who, moreover, showed that, when a body floats in a liquid, its weight is equal to the weight of liquid displaced. Archimedes also considered the theory of the lever, the use of which is illustrated in the sculptures of Assyria and Egypt, two thousand years earlier, for the purpose of moving colossal figures and huge blocks of stone. Nowadays we treat the law of the lever as a matter for experimental verification, and deduce other, more complicated,

Archimedes and the Origins of Mechanics.

results from it. But, with the Greek love of abstract reasoning, Archimedes deduced that law from what he regarded as self-evident propositions. Implicitly, however, the principle of the centre of gravity, which is equivalent to that of the lever, is contained in his proof. Nevertheless, the co-ordination of the law of the lever with ideas which seemed simpler to Archimedes and his contemporaries, was a great step in advance, and the type of all scientific explanation.

When we turn to biology, we find that there too the Greek love of theorizing without facts sometimes led them astray. Anatomy seems first to have attracted the attention of the philosophers, but such knowledge as existed was gained by the dissection of the lower animals only. Considerable progress was made in purely descriptive work. In natural history Aristotle, as we have seen, put together an extensive collection of observations on the animals then known, with some details of their anatomy, probably gained personally by experimental methods.

Medicine.

With regard to the earliest practice of medicine, it is interesting to note that in Homer wounds are treated in a simple, straightforward manner, showing the wholesome tradition of a rational spirit in medicine and surgery among the race of Homeric heroes. But it appears that this view was confined to the one class. Among the rest of the people, as in other southern and eastern lands inhabited by Mediterranean stocks, spells and incantations formed the prevalent mode of treatment. Here again in early times the divergent spirit of the two races makes itself manifest.

As medicine developed, the deductive method so dear to the Greeks took possession, and preconceived views about the nature of man or the origin of life were used as the basis of medical treatment, and doubtless cost many patients their lives. When theorizing was kept within bounds, the art of medicine made rapid progress ; the status of the physician rose with it, and an excellent code of professional life was adopted.

Greek medicine culminated in the school of Hippocrates (450 B.C.), with a theory and practice of the art resembling those which are current to-day, and far in advance of the ideas of any intervening epoch till modern times drew near. Disease was reckoned as a process subject to natural laws. The insistence on minute observation and careful interpretation of symptoms led the way for the foundation of modern clinical medicine, while many diseases were accurately described and appropriate treatment indicated. But it was not till later, probably at Alexandria, under the sway of the Ptolemies, that human anatomy and physiology received a proper basis of ascertained fact by systematic human dissection, an advance which may perhaps be attributed to the opportunities afforded by the Egyptian custom of preparing and embalming the bodies of the dead, as well as to the influence of the old-established medical science of the Nile valley.

The atomic theory, as we have said, was extended, even more speculatively than in physics, to biological questions. Lucretius describes the formation of worlds and all possible forms of life by the " fortuitous concourse of atoms " in the chances of infinite time, while only those systems persisted which were fit for the environment. Here we see a faint forecast of the nebular hypothesis and the Darwinian theory of natural selection.

In classical times, original scientific thought seems to have been confined to the Greeks. Although the
The Failure composition of the inhabitants in the
of Rome. adjoining peninsula, or at any rate in its northern parts, and in the Roman State, was probably of a similar character to that of Greece, *i.e.* an indigenous Mediterranean race, overlain and directed by an incoming people from beyond the Alps, yet the resultant population showed considerable differences in development and achievement. The Romans, with their incomparable instinct for the State, and their transcendent power as administrators and framers of law, had little academic intellectual force, although the numerous compilations that came into being seem to indicate a considerable curiosity about natural objects. Their art, their science, even their medicine, seem to have been borrowed from the Greeks ; and, when Rome became mistress of the world, Greek philosophers and Greek physicians resorted to the banks of the Tiber, without establishing any native schools, there or elsewhere, worthy to succeed those of Athens.

One Roman citizen, born in North Italy, the elder Pliny (23–79 A.D.), is to be remembered for having produced in the thirty-seven books of his *Naturalis Historia* an encyclopædia of the whole science of the period. He placed on record the knowledge and beliefs of a series of forgotten writers and workers of Greece and Rome. Starting from the world, the sky and the stars in space, which he regarded as a kind of pantheistic deity, he passed in review the earth, with terrestrial phenomena, such as earthquakes, and dealt successively with geography, with man, his mental and physical qualities, with animals, birds, trees, agricultural operations, forestry, fruit-growing, wine-making, the nature and use of metals, and the origin and practice of the fine arts. Pliny discourses with equal satisfaction on the natural history of the lion, the unicorn and the phœnix, unable to distinguish between the real and the imaginary, the true, the credible and the impossible. He preserves for us the superstitions of the time, and recounts in all good faith the practice and utility of various forms of magic. But, to his credit, it must not be forgotten that he died a victim to his curiosity in natural knowledge. He was in command of the Roman fleet at the time of the great eruption of Vesuvius which destroyed Pompeii and Herculaneum. He landed in order to watch the development of the upheaval, advanced too far inland, and was overcome and borne down by the storm of falling ashes.

The greatest physician of the Roman period was Galen, born at Pergamos, in Asia Minor, about 130 A.D. Galen reunited the divided schools of medicine, and worked at the dissection of animals. His system of medicine, in opposition to the materialistic views of the atomists, was founded on the Hippocratic theory of four elements, combined with the idea of a spirit pervading all parts of the body. It was for dogmas deduced with great dialectic subtlety from these views, rather than for his experimental observations or practical skill, that Galen became famous, and influenced the practice of medicine for fifteen hundred years.

The chief name which distinguishes the later Græco-Roman Alexandrian school of science is that of Ptolemy, who taught and made observations there between the years 127 and 151 A.D. His great work on astronomy, usually spoken of by its Arabian name of *Almagest*, remained the standard treatise until the days of Kepler and Copernicus ; but, in spite of greater fullness of treatment, it does not alter the theories of celestial phenomena already suggested by Hipparchus. Ptolemy, like his master, improved and developed the science of trigonometry, with the view of basing his observations and their results on the " incontrovertible ways of arithmetic and geometry," and laid down the general principle that, in attempting to explain phenomena, it is necessary to adopt the simplest hypothesis that will co-ordinate the facts.

Ptolemy is perhaps better known as a geographer than as an astronomer, and he exercised an influence in this department of knowledge which only gradually sank into the background under the stimulus of the maritime discoveries of the fifteenth and sixteenth centuries. But, here again, it is difficult to assign the merit of much of the work to the respective shares of Ptolemy himself and his immediate forerunner, Marinus of Tyre, whose writings have not separately survived. Ptolemy undoubtedly placed geography on a secure footing by insisting that correct observations of latitude and longitude must precede any satisfactory attempts at surveying and map-drawing ; but his own materials for carrying out such a design were very inadequate, since there was, indeed, practically no method then by which longitudes could be determined with any accuracy. Nevertheless, put together from information brought by traders and explorers, Ptolemy's maps of the known world, extending from the shores of the Malay peninsula and the coast-line of China to the Straits of Gibraltar and the Fortunate Islands, from Great Britain, the Scandinavian lands, and the Russian steppes to a vague land of lakes at the head-waters of the Nile, retain their interest. His general treatment of the subject is that of an astronomer rather than a geographer, for he makes no attempt to describe climate, natural productions or even the aspects which would now be included under physical geography ; nor does he avail himself, to any large extent, of the descriptions and accounts of lands within the Roman Empire which must have been accessible in the military "itineraries."

Whatever be the cause of the phenomenon, it is clear that, even before the decay of Rome as a political power, science, in common with other forms of thought, had come almost to a standstill. No advance in knowledge was being made, and all that was done was in the direction of writing compendiums and commentaries, chiefly on the Greek philosophers, and in especial on Aristotle, who came to be regarded as the great authority on all questions of scientific theory and even of actual fact.

While Greek learning, receiving no reinforcements, lost itself in the capital and the western territories The Influence of the Roman Empire, it pursued a more of Alexandria. fertile course in the lands and cities of the Eastern Mediterranean, whence it had received its original stimulus. At Alexandria, a school of thought arose, influenced alike by Hellenic culture and Jewish, mingled with Babylonian, tradition. It must be remembered that but a small, and relatively unimportant, number of the Jews returned to Palestine after the close of the Babylonian captivity, and that many of the rest of the nation, establishing themselves as traders in the cities of Asia Minor and the Levant, formed a network of communication, commercial, political and intellectual, throughout the East. Alexandria was the commercial, as Jerusalem remained the religious, centre of this scattered but most influential community. It was by their means that much of the Greek influence retained its vitality, and took part in the formation of thought in the early Christian churches. The ideas of Plato thus passed into the Christian philosophy, and became current in mediæval Europe long before their origin was suspected by the schoolmen, who were afterwards amazed to find the familiar doctrine imbedded in the works of the heathen philosophers. Then, too, when the natural line of descent failed in the fall of Rome, the ancient learning found a way into the schools of mediæval Europe through the Moorish dominions and the Arab conquerors of Spain, who themselves obtained it from Alexandria.

CHAPTER III

THE MEDIÆVAL MIND

The Middle Ages—The Patristic Age—The Dark Ages—The Reconstruction of Europe—The Coming of the North—Arabian Science—The Revival of Learning—The Schoolmen—Thomas Aquinas—Dante—Roger Bacon—The Decay of Scholasticism.

UNTIL recent years, the term " Middle Ages " was applied to the whole long interval of a thousand years between the fall of the ancient civilization and the rise of the Italian Renaissance. But the revival of interest in the history, art and religion of the thirteenth and fourteenth centuries has led to a clear recognition that by that time a new civilization had arisen, and has produced a growing tendency to restrict the name " Mediæval " to the four hundred years which preceded the Renaissance and followed what soon came to be termed specifically the " Dark Ages."

But, to the historian of science, there are advantages in the older classification. It is not till the period of the Renaissance that modern science recovers the knowledge of the ancients, learns to examine it critically, and starts on a path of its own by the help of the new experimental method. Hence the period

from 1100 onward, like the dark ages that preceded
it, is to the historian of science but a time of preparation. The two divisions are part of the same
whole, and may well be treated together, though for
the historian of politics, literature or art they are
distinct and separable. To us, then, the Middle
Ages have their old significance—the ages that come
between the ancient learning and that of the Renaissance—the great dark valley across which mankind
had to struggle down the descent on one side from
the heights of the thought of Greece and the power
of Rome to the slow ascent on the other to the upward
slopes of modern science. We look across the cloud-
filled hollow and see the hills beyond more clearly
than the intervening ground, lit only by " the dim
light of scholasticism and theology."

In order to appreciate the causes which produced
the great failure of the Middle Ages to add to our
The
Patristic Age. stores of natural knowledge, it is necessary
to understand the attitude which char-
acterized the mediæval mind ; to realize that the one
and overwhelming fact is the universal dominance of
the religious motive of salvation. This conception
had been founded on the theology framed by the
early Fathers in terms of Hebrew dogma and Greek
philosophical concepts, and moulded by each succeeding age chiefly as an instrument of controversy to
defend from the attacks of pagan or heretic the
scheme which each regarded as orthodox. Then, in
turn, it is necessary to understand why patristic and
mediæval Christianity was inimical in spirit to secular
learning ; why under its influence philosophy became

the handmaid of theology, and natural science vanished from the earth.

This result was not inherent in the nature of the case ; in the earliest patristic age a different spirit was abroad.

Of all the Fathers of the Church, Origen, one of the earliest, is most nearly akin to the modern mind. He was born at Alexandria, about 185, of parents who were Greek by race and Christian in religion. Origen laid the foundation of textual criticism of both the New and the Old Testaments, applying critical methods even in discussing the historical aspects of Christ's life and person. He laboured to free Christianity from the tyranny of the Jewish-Babylonian doctrines of damnation and hell ; and, by dwelling on the harmony of the Christian faith, as it appeared to him, with the scientific knowledge of the Alexandrian Greeks, in which he was proficient, he exercised a great influence both in predisposing the more intellectual classes of the community towards the new religion and in directing the theological development of the Church. But he did not escape persecution and, probably, imprisonment during his lifetime ; partly for political reasons and partly out of deference to the prejudices of the more ignorant and ascetic monastic communities of the age and district. His teaching was finally anathematized at the Council of Constantinople in 553, and his authority gave way before the narrow and dogmatic bigotry of the later Fathers. Had Origen's views prevailed, the Middle Ages would have recovered far earlier and more completely the spirit of freedom and fearless enquiry in which they were so lamentably wanting. But we must turn to the influence which displaced that of Origen.

As we have seen, the older Greek philosophies were founded frankly on observation of the visible world. With Socrates and Plato the enquiry took a deeper turn, and moved from questions of fact to those of reality, from natural to metaphysical philosophy of an idealistic and mystical tendency: "the Greek mind became entranced with its own creations." To Plato, external facts, whether of nature or of human life and history, only become real when apprehended by the mind. Their true meaning must lie in that aspect of them that accords with the mind's consistent scheme of concepts, for thus alone can the facts be thought of, and therefore thus alone can they be. The inconceivable is in truth the impossible.

Such a philosophy clearly could not foster accurate and unprejudiced observation of nature or of history. The structure of the Universe had to conform to the ideas of Platonic philosophy ; history was in its essence a means of vivifying argument or of pointing illustration.

Aristotle was more interested in the observation of nature than was Plato, though it was in metaphysic and logic rather than in science that his greatest strength lay. But Aristotle's influence, great though it was, gradually ceased to be dominant, and by the sixth century had passed out of fashion for seven hundred years. We shall be brought back to Aristotle at a later stage of our enquiry, but for the present we may pass on.

The philosophy of the Stoics was especially suited to the Roman mind, and must not be overlooked in any estimate of the different streams of thought on which the patristic theologians floated their Ark. For the Stoic, the central reality was the human will. Metaphysics and knowledge of the natural world were only of import as they subserved the ends of his philosophy to be a guide of life and conduct. Stoicism was essentially a scheme of ethics, and physics this time were turned from truthful observation by the preconceptions of morals.

The modes of thought inaugurated by Plato were developed into still more super-rational heights by the Neo-Platonists, whose philosophy was the last product of late paganism. From the Egyptian Plotinus (*d.* 270 A.D.) to Porphyry (*d.* 300) and Iamblicus, philosophy became less and less physical and experimental, and more and more concerned with super-rational ideas. In Plotinus, the highest good was the super-rational contemplation of the Absolute. Plotinus lived in a pure region of "metaphysics warmed with occasional ecstasy." In Porphyry, who had a rational side, and still more in Iamblicus, these views were brought down to practical life, and their application led to growing credulity in magic and sorcery. The soul needs the aid of god, angel, dæmon ; the divine is essentially miraculous, and magic is the path to the divine. Thus Neo-Platonism countenanced and absorbed every popular superstition, every development of sorcery and astrology, and every morbid craving for asceticism, of which a decadent age was prodigal. The life of Iamblicus, as told by a Neo-Platonic biographer, is as full of miracle as Athanasius' contemporary life of St Anthony.

This philosophic atmosphere was mixed with currents

of such Eastern faiths as Mithraism and Manichæism, the latter of which enunciated a dualism of the powers of good and evil, destined to appear again and again in later times. The former, with its rites of initiation and purification, disputed with Christianity the possession of the late Roman Empire. All were non-rational, credulous and mystical.

In such an intellectual environment the light of Christianity came to the Greek and Latin Fathers of the Church. They were products of their age and race, and, however superior in moral power to the contemporary pagans, their intellectual standards were the same, and their philosophy was the philosophy of their time, merely changed in emphasis by the story of the Gospel and the Hebrew dogmas with which the Gospel had become entangled.

Christianity, like Neo-Platonism, was based on the fundamental assumption that the ultimate reality of the Universe was spirit, and it too accepted the super-rational attitude. It has been well said that all men except fools have their irrational sides. But, in the early Fathers, the highest super-rationalism, the love of God and the apprehension of the risen Christ, passed down through every step to the lowest forms of credulity held in common with the pagan populace and the Neo-Platonic philosophers. While Plotinus the pagan and Augustine the Christian lay little stress on divination and magic, or on miracles of the saints, Porphyry and Iamblicus on the one side, and Jerome and Gregory on the other, revel in the dæmonic and the miraculous.

Symbolism, which had shown itself in Neo-Platonism, became extended and developed by the efforts of the

Fathers to co-ordinate the Old Testament with the New, and both with the prevalent modes of thought. What in the Scriptures or in the world of nature conforms to the Christian scheme as interpreted by each Father may be received as fact ; what does not so agree is to be understood in a symbolic sense. Here we have the Christian equivalent to the idealism of the Platonists.

Finally, to understand the patristic, and through it the mediæval, mind, it is necessary to appreciate the overwhelming motive introduced by the Christian conception of sin, the need of redemption, and the fears and hopes of the realistic Jewish-Christian ideas of heaven and hell—the need of mediation to obtain salvation to enter the one, and the fear of damnation in the flames of the other. The pagan world itself had become less self-confident. The decadence of the race had carried mankind far from the bright Greek spirit of life, and the stern Roman joy in home and State. Men were coming to rely more on authority in State and thought, were seized with unrest and vague fears for their safety in this world and the hereafter. The phase recurs at various epochs of history. Even before the ministry of Christ, in Palestine and wherever Jewish influence was felt, eyes were looking for a catastrophic coming of the Kingdom of God, a conception which made the Christian faith of the Apostolic age largely a matter of eschatology, and its rule of life but an *interims Ethik*—a short preparation for the triumphant Second Coming. Perhaps in the patristic age the end of the world had receded a little into the future ; but the day of judgment was still very near, and to each man death was an effective door into the mystery of the next world and the horror of the Shade. Darkness was covering the civilization of the ancient lands, and thick darkness the spirit of mankind, almost obscuring the one transcendent ray of Christ's message of hope and reconciliation.

With such an outlook on life and such a prospect in death, it is no wonder that the Fathers showed small interest in secular knowledge for its own sake. " To discuss the nature and position of the earth," says St Ambrose, " does not help us in our hope of the life to come." With Augustine, God's inscrutable will is the direct and immediate source of all causation. In this atmosphere, natural knowledge was valued only as a means of edification, or as an illustration of the doctrines of the Church or the passages of Scripture. Critical power soon ceased to exist, and anything was believed if it accorded with Scripture as understood by the Fathers. The contemporary knowledge of natural history in the Church, for instance, was soon represented by a second-century compilation called *Physiologus*, or the *Bestiary*, in which the subjects and the accounts of them, originally Christian allegories with imagery taken from the animal world, were frankly ruled by doctrinal considerations. For example, it is stated seriously that the cubs of the lioness are born dead. On the third day the lion roars, and they wake to life. This signifies our Lord's resurrection.

And so with their views of history and biography. The classical historians were always ready to modify their accounts to serve the rhetorical fitness of the situation, and the Church historians carried this tendency to much greater lengths. In their hands, history became a branch of Christian apologetic, and the lives of the saints, the characteristic form of early mediæval literature, became simply a means of edification. Any legend which accorded with the author's conception of the holiness of his subject was received unhesitatingly.

The power of patristic theology was increased indefinitely by the ecclesiastical organization which grew up to enshrine it. And when, with the conversion of the Empire, that organization had the decaying but still overwhelming strength of Roman tradition behind it, it became irresistible. The Roman Empire died, but its soul lived on in the Catholic Church, which took over its framework and its universalist ideals. It was immeasurably easier for the Bishop of Rome to acquire the primacy of the world, and gradually to tighten the bands of uniformity, because even barbarians had come to look on Rome as their metropolis, their Holy City, and the Emperor as their suzerain. Hence probably came the triumph of the more legally minded and intolerant Latin theologians over the less rigid and more philosophical Fathers of the East. The great Council of Nicæa met in 325, with characteristic modesty, to " determine the true nature of God," and thereafter, although St Gregory of Nazianus had declared that he never knew a Council of the Church to end well, a series of Councils defined doctrine more and more accurately, and anathematized more and more forcibly those who differed from their findings.

Thus, although some of the Northern nations were converted to Christianity by Arian teachers, and maintained that form of the faith for many centuries, a great central body of doctrine had been built up before the Dark Ages began, and Rome had become the recognized head of the Catholic Church, with all the universalist tradition of the Empire behind it, and the new power of a compelling religious faith to put new life into its veins, and give it uncontrolled dominion over the minds and consciences of mankind.

Such was the nature of the intellectual position when the last gleams of sunset of the ancient civilization were fading away into the dark night

The Dark Ages.

of the sixth and seventh centuries. And such was the nature of the ideals to which the succeeding ages looked back as they emerged into the feeble light of the new morn, looked back as to a brighter day whose glorious noon culminated in God's crowning revelation by his Son, and whose resplendent eve was illumined by the inspired writings of the Fathers of the Church. It is small wonder that the men of the new time took all that came to them from across the darkness to be endowed with supernatural sanction, and showed no power of critical insight, in which the Fathers themselves had been equally deficient. Almost the only traces of secular learning which survived were the works of Boetius, a Roman of noble birth, who was put to death in 525. Probably a nominal Christian, Boetius was the last to show the true spirit of ancient philosophy, and it is part of the irony of history that he developed into a Christian martyr after his death. He wrote compendiums and commentaries on Aristotle and Plato, and treatises on arithmetic, geometry, and music founded on those of the Greeks. After Boetius the classical spirit vanished from the earth. The schools of philosophy at Athens were closed in 529 by order of the Emperor Justinian.

The break with the past was much more complete than the mere fall of Greece as a civilizing influence and of Rome as a world-power necessarily involved. Not only were Athens and Rome destroyed as political states and social civilizations, but both the race of the Greeks, the artists and philosophers, and the race of the Romans, the lawyers and administrators, had ceased to be. Malaria had made vast tracts of country uninhabitable, swarms of alien immigrants had corrupted the purity of the blood, while the fall in the birth-rate among the nobler and abler stocks and the constant drain of incessant wars and—among the Romans—of foreign administration, had lowered the average quality of the nations. The definite races of Greece and Rome, which had effected such great things in the history of the world, had given place to mongrel, cross-bred populations, composed of incompatible Northern, Mediterranean, African and Semitic elements, fated, by their lack of cohesion, want of common ideals and disregard of statecraft, to inward decay and outward destruction at the hand of the first strong, pure-bred people that encountered them. Thus the overthrow of Rome by Northern Teutons was not essentially a destruction of civilization by barbarians. It was much more the clearing away of a doomed and crumbling ruin, in preparation for future rebuilding.

A new civilization had to be evolved from chaos ; nations with definite ideals and well-marked characteristics had to reform out of the medley of races comprised in the decadent universalist Imperialism ; and those nations had to advance far in the reconstruction of social order and the determination and specialization of intellectual attributes before they could throw off the spell of patristic and Semitic theology, and form a suitable seed-bed for the germination and growth of a new science and scientific philosophy.

Here and there, in the gloom of the Dark Ages, we see tiny plants of knowledge struggling to the light. It is probable that in Italy some of the secular schools maintained their continuity in the large towns throughout the times of turmoil and confusion. But the rise of the monasteries gave the first chance of a secure and leisured life, and, consequently, it is in the cloister that the first signs of the new growth of learning are seen.

In view of the character of the Gospel story, it was impossible for the Fathers of the Church to despise the art of healing as they despised or ignored more speculative secular knowledge. Hence tending the sick remained a Christian duty, and medicine was the earliest science to revive. Monastic medicine was at first a mixture of magic with a faint tincture of ancient science. In the sixth century the Benedictines began the study of compendiums on the works of Hippocrates and Galen, and gradually spread a knowledge of these writings throughout the West.

In the seventh century the first new secular home for learning appears in the schools of Salerno, whence proceeded many compilations founded on Hippocrates and Galen. In the ninth century Salernian physicians were already famous, and the schools flourished till overshadowed by the rise of Arab medicine in the twelfth century. Since Salerno is known to have been first a Greek and then a Roman colony, it is likely that here there existed a direct link between the learning of the ancient and the modern worlds. ·

But it should be noted that countries at a distance from Rome, the centre of the chaos, were among The Recon- the first to show signs of a new and struction of distinctive creative spirit. The literary Europe. and artistic development of Ireland, Scotland and the north of England, beginning with Irish sagas full of poetic extravagances, was quickened by the absorption of Christianity, and thereafter, in the fervour of its missionary spirit, carried back some culture into more southern lands. This northern development culminated in the works of the Anglo-Saxon monk, Bede of Jarrow (673–735), who incorporated into his writings all the knowledge then available in Western Europe. He stands between the Latin commentators Boetius, Cassiodorus, Gregory, and Isidore of Seville, who caught the last direct echoes of classical or patristic learning, and the scholars of the schools of Charlemagne, chief among them Alcuin of York, who carried the tradition into definitely mediæval times. Bede wrote in Latin chiefly for monks ; but a hundred and fifty years later culture had so broadened that Alfred the Great (849–901) translated or caused to be translated into Anglo-Saxon many Latin books. The influence of Latin literature had begun to pass into the native languages.

And indeed by this time mediæval Europe was taking shape. Nations had crystallized out from the mixture of the Romanized Gauls with the vigorous Teutonic tribes that overran the Roman provinces. Northern lands that had never seen the Roman eagles, or from which they had retreated, were developing a culture and even a literature of their own, on which Roman ideas and Roman civilization only acted as external and foreign influences.

We have said that it is to the different branches of the Northern race that modern science is mainly
The Coming due. Even in Italy, when, first of nations,
of the North. the Renaissance touched her, experimental science took its rise in the Northern regions which had been permeated with streams of barbarian blood by the influx of Goth and Lombard. Hence, for our present purpose, it is well to learn what we can of the religion and racial characters of those folk with whose achievements we shall almost exclusively be concerned in our later chapters.

Once more, as in the case of the Greeks, we may see the genius of a people shadowed forth in their religion and mythology. The heathen faith of Gaul and Germany and Britain disappeared on contact with Christianity, leaving well-nigh no trace. But Scandinavia kept its native gods far longer—long enough for poetry, freed from religious prejudices, while deposing the gods, to retain the myths, and to enshrine them for ever in the deathless tales of the North. The Icelandic edda of Snorri Sturlusen was written about 1222 in prose, but it contained quotations in verse from heathen poems, some of which have survived in an older edda. In these poems we catch glimpses of generation behind generation, back into the dim obscurity of the unfathomable past. It is then to the literature of Norway and Iceland that we must look for the purest example of Teutonic mythology.

The most striking feature of the Norse legends is their similarity to those of the Greeks. Odin is a mixture of Zeus and Prometheus; Thor is Zeus in another aspect, with attributes of the Roman Mars superadded. Greeks and Teutons were branches of the one great Northern race, and racial character is reflected in beliefs drawn originally, perhaps, from a common source. But the different environment modified the different branches of the race by natural selection, affected their outlook on life, and moulded the mythology in which that outlook found its most characteristic expression. With the same vigour and freedom and joy in life and adventure as the Greeks, the Norse-men had less sense of soft beauty and grace, and a sterner, somewhat sadder outlook on the problems of the world. The Greek gods had conquered the Titans once for all. Thor and Odin wage perpetual war on the giants of Jotunheim. The Greek gave little thought to the edge of the world dropping down into Tartarus, where sunless Cronos dwells. But the Scandinavian is for ever preoccupied with the icy barrier, closer to his threshold, beyond which lie the gulfs of chaos.

In the early cosmogonies, the world is the body of the giant Hymir, his skull is the heaven, his brains the clouds, his blood the sea. The stars are sparks from the great primordial fiery chaos. Then, as in Greece, there are myths of nature. Thor, the thunder-god, may be taken as the ideal of the Norse races—homely, fearless, steadfast ; with a slow brain perhaps, but judgment sound at the last.

In a later age, when the advance of Christianity threw the shadow of inevitable fate over the heathen gods, the old nature myth of the death of Balder became the symbolic tragedy of all death. Heaven and earth and gods are to pass away, in one great fight on the ramparts of Asgard, when the powers of chaos overwhelm the world and the twilight of the gods descends. The gods fight on the side of reason, hopelessly, knowing that chaos wins, but fearlessly, with steady courage—the glorification of Teutonic self-will and inward freedom of the individual soul. In the Viking age there rose a vision of Balder coming again, and a new heaven and a new earth. But this may be but a reflection of the Christian Apocalypse ; the day of pagan myths was over.

But still the spirit of the North survives in the Sagas, the epics of heroes and heroic kings. Olaf Tryggvason is the ideal Norseman—tall, golden-haired, excelling in sport and terrible in war, eager, glad and kind. All Norway owns his sway, and, at his bidding, turns to the White Christ. The Sagas are direct, convincing ; free from unreality or absurdity ; thrilling with the voice of the sea and the call of battle. Their occasional supernaturalism conjures up visions of trolls, portents and second sight ; it is restrained, almost modern in tone ; quite different in spirit from the tissue of childish marvels and morbid asceticism which characterize the contemporary lives of Southern pseudo-Christian saints, heroes of the bewildered cross-bred Mediterranean peoples.

The heathen faiths had vanished. But the genius

which gave them birth passed over into the new religion
of the Northern race.   At first awed not only by the
light of the Gospel but also by the setting of patristic
theology and decadent Latin culture in which it had
become enshrined, they were long held in free sub-
mission to the form and constitution of the faith they
had accepted.   " If I and my Franks had been there,
we would have avenged Him," said Clovis when
told the story of the Crucifixion ; and, to them, the
Gospel message which moved their hearts was inseparable from the organization and prestige of the
Church, Arian or Roman, in which it reached them.

But, when the first bright charm wore off, famili-
arity with the tyranny of ecclesiastical power wrought
a change in their mental state.   With alternate enthusiastic championship and rough chafing at the
slowly tightening chain of Roman Catholic theology,
their characteristic genius breaks out again and again
in history ; sometimes repressed with fire and sword,
with rack and stake, but triumphant at the last, till
the bonds are broken, till freedom of action and speech
and thought are established in all the lands where the
Teutons hold sway, established by the clash of the
warring parties and sects into which their independent
spirit for ever leads and sometimes betrays them.
Such is the race which laid the foundations and built
most of the superstructure of modern science.

Differences in genius among the various branches
of the race, ancient and modern, may be detected.
The Greek was too self-centred and consequently too
philosophical to grasp the essential spirit of experimental science.   The Roman had too little power of
abstract thought and too much fondness for legal

forms. Thus it was the Teuton, observant, pains-
taking, sure, who created modern science, which still
remains chiefly a Teutonic achievement.

Even the Teutons, by the manifestation of their
typical genius, may be subdivided. For example, in
English history, the Saxons of the south and middle
west have most often excelled in administration and
imaginative literature, and it is to the Angles and
Danes of the east and north-east that we must turn
for the most frequent output of scientific ability.   The
small, dark, indigenous race of the west have been
singularly barren of original scientific achievement.

But, in early mediæval times, all this was still in
the womb of the future ; there was as yet little sign
of the birth of modern science.   The great work of
that age was preparation, the consolidation of nations,
in which afterwards knowledge might develop in
accordance with the special genius of each people.
The traditional Latin culture survived in part in
countries where it had taken root, and permeated
slowly those which successively made their way into
the European comity.   But, for some centuries, little
save this local tradition came over from the past.
For eight hundred years Greek was almost an unknown language in the West, and Greek and Roman
learning but the faint trickle which found a tortuous
path through the late Latin writers of compendiums,
and the interpretations which mediæval minds, saturated with theological preconceptions, put in their
turn upon the views of those Latins.

In Italy, which might have given the best field for
direct contact of old and new, the population had
been corrupted too long and too completely by alien

Eastern and African blood to rally quickly from the poison. The land had been laid waste again and again by invasion, and almost all continuity of knowledge was destroyed, not so much by the overthrow of social conditions, as by the destruction of the minds capable of transmitting it. Thus the natural course by which culture and learning, following the old routes, could reach the new Western nations had silted up, and the stream, when it broke in at last, came by other channels.

It was by indirect paths that most of the re-covered learning of the Greeks reached mediæval Europe. Of these indirect paths perhaps the earliest was the Arabian.

Arabian Science.

A certain amount of Græco-Jewish culture survived for many centuries in Syria and in the neighbourhood of the Persian Gulf, with its chief centres at the Eastern Imperial Court at Constantinople and at Baghdad. It was probably along the Eastern trade routes that the Indian numerals travelled to Europe, where, under Arabian influence, they gradually displaced the clumsy Roman figuring. In Damascus and Baghdad, Greek medicine was still cultivated under Jewish and Christian teachers, and received some accessions of knowledge, especially as regards drugs, from Indian physicians, who began to travel westwards.

Progress was also made in astronomical calculations by Mohamed El Batani (c. 850), who, from his observatory at Antioch, recalculated the precession of the equinoxes and drew up a new set of astronomical tables; while in the eleventh century various observations on solar and lunar eclipses were placed on record by Ibn Junis, who worked near Cairo.

The classical period of Arabian science may be said to date from the tenth century, beginning with the medical work of Rhazes, who practised in Baghdad and compiled many encyclopædic text-books. Fifty years later, the great physician and philosopher Avicenna, a native of Bokhara, wandered from court to court among the rulers of Eastern Asia, vainly seeking some place of settlement where he could find an opening for his talents and carry on his literary and scientific labours in peace and safety. His "Canon," or compendium of medicine, afterwards became the text-book of medical study in the European universities, and up to the year 1650 it was still used in the schools of Louvain and Montpellier.

But any prospect of the establishment of a stable Arabian civilization was put an end to by the internal quarrels of the Mahommedan princes and generals, and by the gradual disintegration and destruction of the gifted, noble and old-established families of Arabia itself. These pure Arabs were the original conquerors, and had provided the necessary governors, soldiers and administrators for the vast empire which had been put together so rapidly. With the accession of the Abbaside Caliphs and the transference of the capital to Baghdad, the real power passed into the hands of a series of Persian, Tartar, and Turcoman ministers and slaves. Gradually the distant provinces, one after another, separated themselves from the weak, overgrown and heterogeneous Empire, re-created their native characters and reasserted their political independence. The countries nearest the capital have never recovered from the prolonged vicissitudes of a line of impotent tyrants, with its fatal insecurity

of universal lawlessness. The apparent quiet which
ultimately settled on them was not that of peace but
of exhaustion. Ease, wealth, trade, literature and
science, all eventually disappeared from the sphere
of influence of the Mahommedan domination.

It was in Spain, the farthest province of the Mahommedan conquest, that the first fruits of the intercourse of Arabian, Jewish and Christian civilizations
became appreciable. For three centuries, from 418
to 711, the West Gothic kingdom had established and
maintained law and order from its capital at Toulouse.
The Sephardim Jews, originally deported from
Palestine to Spain under Titus, had preserved traditions of Alexandrian learning, amassed wealth, and
kept open communications with the East. But
the persecutions of the Jews, which set in after the
West Gothic kings exchanged the more tolerant
Arianism for the orthodox Roman Christianity, pre-
disposed the Jews for a change of governors, an
attitude which helped to turn what might have been
a mere Mahommedan plundering raid into a successful
conquest.

The tolerance of thought accorded by the Arabs,
as long as their supremacy remained unquestioned,
led the way for the establishment of schools and colleges,
which, however, owed their continued existence, not
to the general intelligence of the people as a whole,
but to the occasional and spasmodic patronage of a
liberal-minded or free-thinking ruler.

The course of Arabian philosophy developed on
much the same lines as that of the Christian schools,
which followed it a hundred years later. There was
the same attempt to harmonize the sacred literature

of the nation with the teachings of Greek philosophy, and an analogous contest between those who relied on reason and rational authority and those who put their faith in revelation, denying the validity of human reason in matters of faith.

The chief fame of this Spanish-Arabian school of thought is due to the work of Averrhoes, who was born at Cordova in 1126. While showing a profound reverence for the teachings of Aristotle, Averrhoes nevertheless introduced a new conception into the relations between religion and philosophy. According to him, religion was not a branch of knowledge that could be reduced to propositions and systems of dogma, but a personal and inward power, distinct from the generalities of " demonstrative " or experimental science. Theology, the mixture of the two, he regarded as an unmixed source of evil to both, fostering, on the one hand, a false impression of the hostility between religion and philosophy, and, on the other, corrupting religion by a pseudo-science.

It is not surprising that the teaching of Averrhoes came into fierce conflict with that of the orthodox theologians, but, in spite of unrelenting opposition, especially from the great Dominican school of thought, his words fell upon willing ears. By the thirteenth century, Averrhoes had become a recognized authority in the universities of North Italy, Paris and Oxford, worthy to be placed, according to Roger Bacon and Duns Scotus, by the side of Aristotle as a master of the science of proof. That such a position should have been accorded to a teacher who was not only of non-European origin, but also of alien and antagonistic religion, indicates the approach of a time when neither creed nor nationality should form an insuperable barrier in the world of thought.

In the Europe which received and slowly absorbed this Arabian stream of knowledge, the apparatus of The Revival learning had made considerable pro-of Learning. gress. As we have seen, a home of secular studies, and especially of medicine, had existed since the seventh century at Salerno, and, in Northern Europe, the encouragement bestowed on scholars by Charlemagne and Alfred had given an impetus to teaching generally. The monastic and cathedral schools, hitherto in most places the only educational institutions, were soon found insufficient to meet the growing needs, and new secular schools began to assume their later form as definite universities.

A revival of legal studies took place in Bologna about the year 1000, and, in the twelfth century, schools of medicine and philosophy were added to that of law. A students' guild, or Universitas, was formed for the mutual protection, at first of the foreign students, who were at the mercy of the inhabitants, and later of all students, whether native or foreign. These guilds hired their own teachers, and thus the University of Bologna, even in its after life, continued to be a students' university, in which the governing power was in the hands of the learners.

On the other hand, a school existed at Paris in the first decade of the twelfth century, conducted by teachers of dialectic, and shortly afterwards a community or Universitas of teachers set the constitutional model to all the subsequent universities of northern continental Europe and England. Thus it

6

is, that at Oxford and Cambridge the governing power has always rested with the teachers and not with the students, as at Bologna and to some extent in Scotland, where the election of the Rector is one of the last traces of undergraduate control.

As early as the Carolingian period, the academic subjects of study had settled down into an elementary trivium, comprising grammar, rhetoric and dialectic, subjects which dealt with words, and a more advanced quadrivium—music, arithmetic, geometry and astronomy, which were supposed, at all events, to deal with things. Music contained a half-mystical doctrine of numbers, geometry merely a series of Euclid's propositions without the proofs, while arithmetic and astronomy were esteemed chiefly because they taught the means of fixing the date of Easter. All were studied as a preparation for the study of the sacred science theology. All through the Middle Ages, this division of the subjects of study held good for the elements of academic learning, and, with the growing interest in philosophy, that study was merely superadded to the simpler logical dialectic.

The old controversy between Plato and Aristotle on the nature of the idea of universals found its way into the commentaries of Porphyry and Boetius, and so reached the mediæval mind. What is the meaning of the power of classification ? Are the individuals the only realities, the universals existing merely in and with the objects of sense as their essence, as taught by Aristotle, or even only as names or mental concepts as the extreme Nominalists held ? Or had the ideas or universals an existence apart from the phenomena or the isolated beings, as in Plato's philosophy ?

In the ninth century the mystical theory of Erigena, a native of the British Isles and disciple of Origen, a theory based on the idea of the divine as the only reality, contained a fusion of Platonic and Aristotelian views, and the discussion only became acute later. In the eleventh century pure nominalism appears in Berengarius of Tours (999–1088), and later in Roscellinus (d. c. 1125), who, holding the sole reality of the individual, reached a tritheistic conception of the Trinity, and at once crystallized, especially in William of Champeaux and Anselm, the opposing realism, and established it as the orthodox view for several centuries. But its inherent difficulties led to many varieties of the theory, and an interminable discussion waged in the schools and employed all the philosophic interests of the scholastic dialecticians for two hundred years. Abelard, a Breton noble by birth, attacked his master, William of Champeaux, and himself taught a modified Aristotelian realism verging on a nominalism not so consistent as that of Roscellinus. In Abelard the doctrine of the Trinity was reduced to the conception of three aspects of the Divine Being. Abelard showed signs of rationalism, such as the pregnant statements that "doubt is the road to enquiry," that "by enquiry we perceive the truth," and that "it is necessary to understand in order to believe," a saying that may well be compared with the "Credo quia impossibile" of the patristic Tertullian, and the "Credo ut intelligam" of Anselm. Abelard was called to account by St Bernard, who held in abhorrence the wisdom of this world, and did much to fan the growing spirit of ecclesiastical suspicion which saw heresy everywhere. In this controversy

we find the first notable instance of that conflict of
ecclesiastical authority with the growing demands of
the human reason which became the dominant intellectual feature of the coming centuries. But for a
time the speculative spirit was exhausted, and the
middle of the twelfth century marks the beginning of
a pause of fifty years in logical and philosophical
dialectic, and the return to a passing interest in
classical literature, an interest centred in John of
Salisbury and his school at Chartres.

If the intellectual task of the Dark Ages was to save
what it could out of the wreck of the ancient learning,

The
Schoolmen.
that of the first succeeding centuries was
to master and absorb what was recovered.
As we have seen, the intellectual achievement of
the mediæval period was the welding together of
the remains of the ancient classical knowledge and
the Christian faith, as interpreted by the early
Fathers, in the minds of new races to whom both
were foreign. From the eighth century onward we
may watch this process at work, and there the
constructive period of the Middle Ages may be said
to begin.

By the middle of the twelfth century the dual
heritage from the past had been surveyed and mapped
out, absorbed and transformed by the Teutonic mind,
where it lay awhile, awaiting an opportunity to
germinate. Here we have the culmination of mediæval
appreciation of classical literature. None of the more
advanced works of Aristotle were known in a complete
form ; thus no scientific and unliterary source had
come to hand to disturb the literary outlook of those

scholars who cared for the classics as a bypath of study, or as a means of understanding better the language of Scripture and the writings of the Fathers. The predominant theological attitude was still Platonic and Augustinian, idealistic and mystical rather than rationalistically philosophical.

But in the thirteenth century a great change of outlook took place, coincident with and inseparable from the humanizing movement, associated with the coming and establishment of the Friars. Between 1210 and 1225 the complete works of Aristotle were recovered and rendered into Latin, first from imperfect Arab versions, and then by direct translations from the Greek. In this latter work one of the foremost scholars was Robert Grosseteste, patron of the new uncloistered mendicant orders, the great Chancellor of Oxford, and Bishop of Lincoln, who invited Greeks to England and imported Greek books, while his pupil, Roger Bacon, a Franciscan friar, wrote a grammar of the Greek language. Their aim was not literary but philosophic—to unlock the original tongue of Scripture and of Aristotle. Their ultimate though distant achievement was to set free the Western mind from the dialectic subleties of the schoolmen and the Fathers of the Church.

Aristotle opened a new world of thought to the mediæval mind. His attitude—at once more rational and more scientific—was quite different from the Neo-Platonism which hitherto had chiefly represented ancient philosophy. His range of knowledge both in philosophy and in the science of nature was far greater than anything then available. It was a heavy task to absorb and adapt the new material to Middle-Age Christian thought, and the work was not effected without misgiving.

It is difficult now to understand the immense influence that Aristotle came to possess over the schoolmen of the thirteenth century. It is perhaps easier to sympathize in the detestation with which some of the harbingers of the Renaissance regarded that philosopher and all his works, the loathing for an authority that checked their eager minds at every turn. But to the mediæval churchman, convinced of the intellectual supremacy of his Church as the recipient of all revealed knowledge and the only begetter of true thought, it was marvellous to find the cherished doctrines enshrined in the work of men who lived and died centuries before the revelation of their Lord. It never entered their consciousness that the recovered volumes were actually the source of much of the inspiration of the holy Fathers. We may realize some of their confusion in Dante's treatment of his pagan master Virgil, who ranks highest amongst mortal men, and is competent to guide the Christian pilgrim far on his way and need only turn back at the threshold of the supreme accomplishment.

At first the Arabian channels by which Aristotle reached the West mixed his philosophy with some Averrhoist leanings, and mystical heresies were the result. Aristotle's works were condemned by the provincial Council which met at Paris in 1209, and this prohibition was renewed later. But the new knowledge was irresistible, and in 1255 the University of Paris formally placed Aristotle's works upon the list of books to be studied.

The interpretation of Aristotelian thought in terms of Christian dogma, and its incorporation into the Thomas Aquinas. official philosophy of the Roman Church, were chiefly the work of one man, Thomas Aquinas, who systematized and reduced to order the compendious writings of his master, Albertus Magnus.

The importance of Aquinas in the history of philosophy and of the origins of science is great. He delayed for many years the liberation of scientific thought from the trammels of theology, and to his indirect influence is chiefly due the obscurantist attitude of Rome towards the initial development of modern science at the Renaissance. Even as late as 1879, Pope Leo XIII. directed Roman Catholic teachers to found their philosophy on that of Aquinas. Hence, for our purpose, it is necessary clearly to understand the essence of the man and his work. His task was to reconcile faith and reason—a task admirable in intention, boldly and skilfully executed, yet disastrous for mankind in that it was at the least premature, and therefore resulted in petrifying for a time both theology and scientific philosophy into one block of what soon became incompatible elements.

Thomas was born about 1225 in Southern Italy, but his ancestry was of mingled Swabian and Norman princely blood, and his father was Count of Aquinum. At the age of eighteen he joined the Dominican order. He studied at Cologne under Albertus Magnus, a native of Swabia, taught at Paris and Rome, and, after a life of incessant activity, died in 1274.

His greatest works, the *Summa Theologiæ* and the *Summa Philosophica contra Gentiles*—the setting forth of Christian knowledge for the ignorant—contain a complete and consistent body of doctrine, in which the statements of Scripture and the physics and metaphysics of Aristotle equally find a place. There are two sources of knowledge, the mysteries of the Christian faith as transmitted by the channels of Scripture, of the Fathers and of Church tradition, and the truths of human reason—not the fallible individual reason, but the fount of natural truth of which the chief channels were Plato and Aristotle. The two sources cannot be contradictory, since they flow from God as the one source. Hence philosophy and theology must be compatible, and a *Summa Theologiæ* should contain the whole of knowledge. Even the highest mysteries of the Trinity and the Incarnation, though they cannot be proved by reason, can be examined and apprehended thereby. Hence these mysteries also are welded into the Thomist system. Throughout, Aquinas' interests are intellectual. Perfect beatitude of any created intellectual nature lies in the action of the intelligence directed to the contemplation of God. Faith and revelation are belief in a proposition and a presentment of truth.

With this outlook, Thomas proceeds to give an account of the whole of knowledge in mingled terms of Scripture, the Fathers and Aristotle. The existence and attributes of God, of angels and of men are successively passed in review. The whole of creation is a *processio*, a going out of all creatures from God ; ideas are prototypal forms existing in the divine mind, and the meaning of individuation is to be sought in the existence of determinate matter. This solution of the old problem of universals involved the obvious difficulty of explaining the individual existence of angels and of the disembodied human soul. Averrhoes had held that individuality was extinguished at death, and to avoid this conclusion Aquinas assumed that such spirits possessed a principle of individuation in themselves. They are immaterial forms since they can grasp the universal, and the soul unit is not merely the rational spirit of Aristotle, but possesses sensitive functions which depend on bodily perceptions in life, but are in essence independent of them.

But, for us, more important is the fact that the whole of the scheme of Aquinas was framed in accordance with the Ptolemaic astronomy and saturated with geocentric ideas, with the view that all motion implied a continual exertion of force, and with the whole spirit of Aristotelian science. From these premisses Aquinas deduced results in accordance with the theology of his age, such as : " *Movetur igitur corpus celeste a substantia intellectuali.*" The deductions being thus regarded as verified, the premisses became strengthened, and doubt thrown on them was soon regarded as inconsistent with the Christian conclusions with which in Thomas' scheme they were connected. The whole of knowledge was welded with dogmatic theology into one rigid structure, the parts of which were believed to be interdependent, so that an attack on Aristotelian science became an attack on the Christian faith.

If Aquinas is the summit of the intellectual side of mediævalism, Dante represents its highest poetic achievement. In Dante the whole basis is scholastic and especially Thomist, with the characteristic mediæval blending of classic and

Dante.

patristic ideas, and symbolic interpretation of Scripture, history and nature.   In the *Paradiso* especially
the structure of the physical and astronomical universe
and of animated creation is set forth in scholastic and
Aristotelian terms.   Yet on this basis the genius of the
poet rises through intellectualist theology to the loftier
insight of direct contemplation, and reaches forward
to the more mystical ideas of the coming age.

The thirteenth century saw the triumphant and
applauded work of Thomas Aquinas, the greatest

Roger
Bacon.

exponent of the scholastic philosophy, and
it saw also the tragic life of Roger Bacon,
the only man throughout the Middle Ages, as far as
records have reached us, who approaches in spirit the
men of science of the Renaissance.   The tragedy of
Bacon's life was as much internal as external, as much
due to the necessary limitations of his modes of
thought in the existing intellectual environment, as
to the persecutions of ecclesiastical authorities.

Roger Bacon was born about the year 1210, near
Ilchester, in the Somerset fens.   His family seem to
have been people of position and considerable wealth.
Roger studied at Oxford, where he came under the
influence in especial of two men, both East Anglians,
Adam Marsh, the mathematician, and Robert Grosseteste, Chancellor of Oxford, and afterwards Bishop of
Lincoln.   " But one alone knows the sciences, the
Bishop of Lincoln," said Bacon ; and again, " In our
days Lord Robert, lately Bishop of Lincoln, and
brother Adam Marsh were perfect in all knowledge."

Grosseteste seems to have been the first in England,
perhaps in Western Europe, to invite Greeks to come

from the East as instructors in the ancient form of their language, which was still read at Constantinople. Bacon himself was equally impressed with the importance of the study of the original language of Aristotle and the New Testament, and put together a book on Greek grammar. He is never tired of insisting that the prevailing ignorance of the original tongues was the cause of that failure in theology and philosophy of which he accused the doctors of the time. In anticipation of modern textual criticism, he points out how the Fathers adapted their translations to the prejudices of their age ; and how subsequent corruptions have crept in through carelessness and ignorance, or by that tampering with the text which had gone on, especially among the Dominicans. Bacon himself was a Franciscan, be it observed.

But that which marks Bacon out from among the other philosophers of his time—indeed of the whole period of the Middle Ages—is his clear understanding that experimental methods alone give certainty in science. This was a revolutionary change of mental attitude, only to be appreciated after a course of study of the other writings of his day. Instead of taking the facts and inferences of natural knowledge from Scripture, the Fathers or even Aristotle, Bacon told the world that the only way to reach truth was to observe and to experiment—again an anticipation, this time of the famous doctrine of his more renowned namesake, Lord Chancellor Francis Bacon, who lived three hundred and fifty years later.

And the Franciscan friar, at all events, seems to have practised his own theories, for he says he spent two thousand pounds, an enormous sum for those days,

on his researches.   Naturally, he was accused of magic
—in which, indeed, he seems to have believed.   After
spending some years in Paris, where he was made a
doctor, he returned to Oxford.   But the growing
suspicions of his labours caused him to be sent back
to Paris for more strict supervision by his Order,
and he seems to have been forbidden to write or to
teach his doctrines.   But now came the chance of
Bacon's life.

Guy de Foulques, an open-minded jurist, warrior
and statesman, who had become interested in Bacon's
work at Paris, was elected Pope, taking the name of
Clement IV.   Bacon wrote to him, and in reply,
Clement sent a letter to " Dilecto filio, Fratri Rogerio
dicto Bacon, Ordinis Fratrum Minorum," commanding
him, notwithstanding the prohibition of any prelate
or the constitution of his Order, to write out the work
which he had been formerly asked for.   For some
unknown reason, the Pope added an injunction of
secrecy, which added to Brother Roger's difficulties.
As a friar he was now pledged to poverty, but, by
borrowing from friends, he got together enough to
provide materials, and, in some fifteen or eighteen
months, despatched in 1267 three books to Clement :
an *Opus Majus*, containing his views at length, an
Opus Minus, or epitome, and an *Opus Tertium*, sent
after the others, for fear of miscarriage.   It is from
these books that we chiefly know the work of Roger
Bacon.

Clement died soon after, and Bacon, deprived of his
protection, suffered without redress a sentence of
imprisonment passed in 1277 by Jerome of Ascoli,
General of the Franciscans, who became Pope

Nicholas IV. It is probable that Bacon was not released till the death of Nicholas in 1292. In that year he wrote a tract called *Compendium Theologiæ*, and we hear no more of the great Friar.

Bacon, for all his change of outlook, accepted most of the mediæval attitude of mind. No man can do more than advance a little way in front of the ranks of that contemporary army of thought to which, whether he will or no, he belongs. Bacon accepted the absolute authority of Scripture, could the pure text be recovered, and the entire frame of dogmatic theology in which Christianity was presented to that age. A more hampering preconception was his agreement with the scholastic view, which in other ways he assails so vehemently, that the end of all science and philosophy was to elucidate and adorn their queen theology. Hence came the confusion and inconsistencies which at every turn are seen in his writings, mixed with originality and insight far beyond his time, or, indeed, of the next three centuries. Struggle as he might, he never cast off the shackles of the theological mind; it mars his work, as the persecution of his Order marred his life.

It is one of the signs of Bacon's greatness that he realized the importance of a study of mathematics as a basis for other sciences. He tells us that mathematics and optics, which he calls perspective, were understood by Robert of Lincoln, and must underlie other studies. Mathematical tables and instruments are necessary, though costly and liable to destruction. He points out the errors of the calendar, and calculates that it had gained one day in excess for each 130 years.

He anticipates some recent speculations in suggesting that climate may influence the laws and social institutions of men.   He gives a long description of the countries of the known world.

He seems especially to have been interested in optics, and describes the laws of reflection and the general phenomena of refraction.   He understands mirrors and lenses, and describes a telescope, though he does not appear to have made one.   As an example of the use of inductive reasoning he gives a theory of the rainbow.

He describes many mechanical inventions, some actually known to him and some as possibilities of the future—among the latter, mechanically driven ships and carriages, and flying machines.   He considers magic mirrors, burning-glasses, gunpowder, Greek fire, the magnet, artificial gold, the philosopher's stone, all in a confused mixture of fact, prediction and credulity.

In trying to appraise the value of Bacon's work we must not forget that his fame would have rested on mere popular tradition of his magic had not Pope Clement commanded him to write his books.   Doubtless others besides Bacon were touched by the same interests but have failed to leave direct traces.   Indeed, reflections of the work of such men are found in Bacon's writings.   " There are only two perfect mathematicians," he writes, " Master John of London, and Master Peter de Maharn-Curia, a Picard."   Master Peter recurs when Bacon is dealing with experiment.

There is one science, Bacon says, more perfect than others, which indeed is needed to verify the others, the science of experiment, surpassing the sciences dependent on argument, since they do not bring certainty, however strong the reasoning, unless experiment be added to test their conclusions. This method of experiment no one understands save Master Peter alone ; he, indeed, is *dominus experimentorum*, but cares not to publish his work, nor for the honours and riches it would bring—nor perhaps, one might venture to surmise, for the risk of an ecclesiastical prison it also might possibly entail.

But, whatever be the truth about these phantom figures which flit across Bacon's pages, it is clear that Friar Roger himself was in spirit a true man of science, born out of due time and chafing unconsciously against the limitations of his own restricted outlook, no less than against the external obstacles at which he rails so openly and so often ; a true harbinger of the ages of experiment, of whom Somerset, Oxford and England may well be proud.

Roger Bacon's criticism of the scholasticism of Aquinas, though effective from the modern point of The Decay of view, was out of harmony with the pre-Scholasticism. vailing spirit of the time, and consequently produced little effect.

Much more damaging were the philosophic attacks which began towards the close of the thirteenth century. Duns Scotus (*c.* 1265–1308), who taught at Oxford and Paris, enlarged the theological ground which even Aquinas had reserved as beyond the demonstration of reason. He based the leading Christian doctrines on the arbitrary Will of God, and took free will as the primary attribute of man, placing it high above reason. Here is the beginning of the revolt against the union of philosophy and religion which the scholastics sought, and which his age believed Thomas Aquinas to have finally and conclusively achieved. A revival of dualism appears, essentially unsatisfying and incomplete, yet necessary as a stage of progress in order that philosophy may be set free from its bondage as the " handmaid of theology," free in fertile union with experiment to give birth to science.

The process went much further in William of Occam, a native of Surrey (d. 1347), who denied that any theological doctrines were rationally demonstrable, and showed the irrational nature of many of the doctrines of the Church. He attacked the extreme theory of papal supremacy, and headed a revolt of Franciscans against the control of Pope John XXII. His writings in defence of this action led to his trial for heresy and imprisonment at Avignon. But he escaped and sought protection of the Emperor Louis of Bavaria, and aided him in his long controversy with the Pope.

This principle of the twofold nature of truth, the acceptance by faith of the doctrines of the Church, and the examination by reason of the subjects of philosophy, was bound up with the revival of nominalism, the belief in the sole reality of individual things, and the reference of universal ideas to the rank of mere names or mental concepts. Stress was laid on the objects of immediate sense-perception, in a spirit that distrusted abstractions, and made eventually for direct observation and experiment, for inductive research.

The new nominalism was opposed and banned by

the Church, and Occam's writings condemned by the University of Paris, which tried to impose realism as late as 1473.   But irresistibly the doctrine spread, and a few years later the show of resistance was abandoned.

Occam's final and most decisive triumph came fifty years later, when Martin Luther (1483–1546) pored over his works in the recently founded University of Wittenberg, and based the new teaching on the nature of the sacrament of the Eucharist on theories drawn from the writings of his " lieber Meister." But, if henceforward philosophy was more able to press home its enquiries free from the need of reaching conclusions foreordained by theology, on the other side religion was for a time detached from rationalism, and given an interval for the development of its no less important emotional and mystical sides.   Hence the fourteenth and fifteenth centuries saw the growth of Northern mysticism, especially in Germany, and the appearance of many types of religious consciousness still existent and of value.

The task of the Middle Ages was accomplished ; the ground prepared for the Renaissance, with humanism, art, practical discovery and the beginnings of natural science as its characteristic glories.   With the decay and death of scholasticism we turn a new page in the history of the world.

For the historian of science mediæval times are the seed-bed of modern growth.   They are a period of surpassing interest, but they have no science of their own till Roger Bacon stands roughly chafing at his limitations external and internal, shaking the door which shuts him from his natural home in later

7

centuries. The interest for us of mediæval thought
is the interest of tracing the changing attitude of the
human mind, as it passes through states where science
would have been impossible, to a condition in which
its rise follows naturally from the philosophic environment.

In a sense we have seen the worst aspect of the
Middle Ages. They are weakest on the intellectual
side, and on that side weakest of all in the special
department of thought necessary for scientific philosophy. We have but glanced in passing at their work
of forming and consolidating the nations of Europe.
We have not touched on their wonderful achievements
in literature and art. The romances of chivalry are
outside our ken. Dante's *Divine Comedy* has for
us little significance, save as the enshrinement in
poetry of the scheme of Thomas Aquinas. The glories
of cathedral architecture are to us but illustrations
of the growth of the builder's art. Even mediæval
religion, which on its unlucky philosophic side has
concerned us nearly, in its essence does not touch
our enquiry. Its saving faith in its divine Founder,
its spirit of humble reverence and love for all mankind
and all creation, its message of salvation to suffering
humanity, are hid from our eyes. We meet St Bernard
the suspicious Inquisitor, but St Francis of Assisi,
loving, joyous, simple-hearted, does not appear in our
pages.

CHAPTER IV

THE RENAISSANCE AND ITS ACHIEVEMENT

The Origins—Leonardo da Vinci—Copernicus—Bruno—Galileo—Gilbert of Colchester—Francis Bacon—The Renaissance in France—Descartes —Pascal—Tycho Brahe and Kepler—Scientific Academies—Newton —Medical Science during the Renaissance—William Harvey—The Rise of Chemistry — Voyages of Discovery — Botany and Natural History — Spontaneous Generation — Witchcraft — The Christian Platonists—Summary.

AFTER the thirteenth century there was a distinct check in the intellectual development of Western Europe. The economic and social confusion caused by the Black Death and the ravages of the Hundred Years' War gave little hope of settled life and quiet study, while in all countries the vast influx of the best men and women of the time into the celibate monastic orders must have produced a disastrous effect on the stock of hereditary ability in Christian lands.

Nevertheless, there was a continual process of change in the mental outlook of mankind, and we may trace, throughout this period of transition, the various streams of thought which, when they met in full vigour, formed the irresistible torrent of the Renaissance. The loosening of the universal grip of scholastic ideas by the solvent influence of the philosophy of

Duns Scotus and William of Occam has already been indicated, and the flight of Occam from a Papal prison to the protection of Louis of Bavaria marks a significant revolt against the power of the Church, a setting up of the rights of nationalities against the universalist tradition of ecclesiastical authority.

The spirit of the Renaissance made its first appearance in Italy, then slowly recovering from the devastation of earlier times. Perhaps the remains of Roman architecture made it easier for the love of the classics to return. A vigorous Northern race had colonized North Italy. They formed the upper class, and were not yet exterminated by the local wars between Italian states then and afterwards so fatal to the nobility. But other lands had purer Northern blood, and the chief Italian advantage in the growth of learning must be sought elsewhere. The clue is given by Salimbene of Parma, a thirteenth-century Franciscan, who remarks on the difference between Italy and other countries in one significant particular. While north of the Alps only the townspeople dwelt in the towns, and the "knights and noble ladies" lived on their estates and superintended the management of their lands in feudal isolation, in Italy the upper class possessed houses in the cities and there passed most of their time.

Now while the residence in the country of its natural leaders makes for a healthy and stable political and social life, in an age of slow communication it gives little chance for that contact of mind with mind which leads to creation and culture. The city life of the leisured and intelligent class in Northern Italy gave an ideal environment for the birth of the Renaissance.

The Renaissance was very far from being an exclusively literary movement. Many other influences conspired to produce an unprecedented intellectual ferment. But one of the most important elements was literary, and with that we may begin our survey.

The harbinger of the literary Renaissance was Petrarch (1304–1374), in whom we see a spirit quite different from the scholastic mediævalism which underlay the poetry of Dante. Petrarch was the first scholar who tried to restore a taste for good classical Latin in place of the dog-Latin of the schoolmen, and, an even more important fact, to recover the true spirit of classical thought, with its claim of absolute liberty for the reason.

Petrarch sang before his time, but, by the opening years of the fifteenth century, a growing interest in classical, and especially in Greek, literature attracted many Greeks from the East, who, from their knowledge of the modern tongue, were able to teach its ancient prototype. The capture of Constantinople by the Turks in 1453 hastened this process, and led to the arrival of many competent teachers, who brought manuscripts with them to their new homes. The search for manuscripts became a keen delight. The monastic and cathedral libraries of Italy and then of Northern Europe were ransacked, while merchant princes with agencies in the Levant used all their resources to procure the copies of Greek writers which had remained for centuries hidden in the East or had been scattered on the fall of Constantinople. In this way, the language of ancient philosophy and science became familiar to Western scholars after a lapse of some eight or nine hundred years.

But even of more value than the language was the free spirit of enquiry it enshrined, and the impulse toward study of all kinds that "humane letters" gave once more to Europe after centuries of gloomy mediævalism. Though the stress laid on the teachings of the Greek philosophers had dangers of its own, the humanists prepared the way for the coming revival of science, and to them is due the chief part in the enlargement of the mental horizon which alone made science possible. Without them, men with scientific minds would never have thrown off their internal fetters of theological preconception ; without them, external obstacles might have proved insurmountable. We seem to see the essential inward union of science, waiting to be born, and religion, struggling to be free, in such men as the Florentine patriot and martyr, Girolamo Savonarola (1453–1498), who, grandson of a skilful Paduan physician, was himself trained for the medical profession, and the high-born Pico del Mirandola (1463–1494), the disciple of Savonarola, whose eager studies and learned wanderings resulted in a great mystical exposition of the process of creation and led him to pronounce the startling opinion that Origen was more probably saved than damned by the verdict of the Council of Constantinople.

For one bright interval, culminating with Pope Leo X. (1513–1521), the Vatican itself was a vitalizing centre of the ancient culture. But the capture of Rome by the Imperial troops in 1527 broke up this new Roman world of intellectual and artistic life, and soon afterwards the Papacy, by reversing its previous policy of liberal guidance, and opposing blindly when it was no longer able to understand or to control, became the greatest obstacle in the way of modern learning of all kinds. Hence the birth of modern science took place amid the throes of persecution, and won its greatest triumphs only in countries that were able to free themselves both from the dead hand of ecclesiastical authority and from the grievous blight of religious warfare.

While the only method of reduplicating books consisted in the laborious process of manual copying on the somewhat costly and troublesome material parchment, the possession of a library was within the reach of but few individuals and not many institutions. The introduction into Europe of the art of making paper followed the later Crusades, and, about a century afterwards, the invention of movable type transformed the old attempts at printing with fixed moulds into a practically useful art, and slowly put books into the hands of all men.

Simultaneously, a renewed ardour of geographical discovery was increasing the area of the earth known to Europe. The Portuguese were the first in the field, and, under the inspiration of Dom Henrique, Prince Henry the Navigator, had pushed south along the western shores of Africa, first on a mission to convert the heathen, and then on the open search for slaves and gold. Their success drove others to emulate them. The Greek theory of the sphericity of the earth was revived and became a general belief. It led to the obvious idea, which indeed the Greeks themselves had propounded, that, by sailing westward into the Atlantic Ocean, the eastern shores of Asia might be reached, and the rich trade of Cathay brought direct by sea to Europe. After many failures, there came the man and the hour. Columbus, born at Cogoletto, on the Ligurian coast of North Italy, landed on the Bahamas on October 12th, 1492, and claimed the lands he had discovered for the crown of Spain. Twenty-four years later Magalhaes' vessel returned after a three years' voyage, having demonstrated the spherical nature of the earth by the convincing proof of circumnavigation.

The wider mental outlook produced by these great voyages of discovery, though the most direct, was not their only effect on the human mind. As the trade with the new lands expanded, the material resources of Europe increased, and the stimulating stream of gold produced an economic development unprecedented in former ages. Wealth, and the leisure for intellectual pursuits which wealth gives, thus spread into far wider circles than were possible with the slender resources of mediæval times. It is worthy of note in the history of mankind that the two periods in which the most surprising intellectual developments are found—the crowning age of Greece and the century of the Renaissance—are each times of expansion geographically and of increased opportunities for a leisured life—in Greece founded on a basis of slavery, and in Europe produced by the wealth of the Indies. In Greece, the age of intellectual triumph was followed all too soon by political disintegration, while the numbers of the nation were always comparatively small. In modern times the Renaissance ushered in a period of four hundred years during which the power of the nations of Europe increased enormously on the whole, while the great growth in population steadily put more and more able men at the service of learning, till the enquirers surpassed in number almost immeasurably the philosophers of Greece. It is perhaps well to bear in mind this last fact when exalting modern achievements in the realms of science.

But, when we have traced what we know of the different tendencies which combined make up the Renaissance, and given due weight to them all, we cannot but feel that the attempt to explain by obvious causes the amazing change of mental attitude produced in so short a time is not wholly successful. As Bishop Creighton said, " After marshalling all the forces and ideas which were at work to produce " this change, the observer " still feels that there was behind all these an animating spirit which he can but most imperfectly catch, whose power blended all else together and gave a sudden cohesion to the whole. This modern spirit formed itself with surprising rapidity, and we cannot fully explain the process."

Leonardo da Vinci.

We must, however, remember that we possess records of but a tithe of the intellectual efforts of the time. Few men then put their thoughts on paper, and of the writings of those few not all have reached us. In Italian city life, knowledge, and the change of outlook which knowledge brings, must have passed from man to man more by word of mouth than by writings. Even in our days, the influence of a great personality by direct action on others is not small, and in the fourteenth and fifteenth centuries it must have been greater in proportion.

The full greatness of one such personality is only now coming to light as the inchoate manuscript note-books of that tremendous universal genius Leonardo da Vinci are transcribed and given to the world. Leonardo may have meant to collect and systematize his notes, and publish them as books. If so, he never lived to carry out his intention. Hence his power as a philosopher and a man of science has hitherto been overshadowed by his fame as an artist.

Leonardo was the natural son of Ser Piero da Vinci, a lawyer of great vigour and some eminence, and was born at Vinci, between Florence and Pisa, in 1452. His mother was a peasant girl named Catarina. He was educated by his father, entered successively the service of the courts of Florence, Milan and Rome, and died in 1519 in France, the servant and friend of Francis I. In early life he showed the remarkable qualities which have impressed both contemporaries and after ages as placing him in a class apart from the rest of mankind. Beauty of person and charm of manner did but adorn and increase the power of mind and force of character which took all knowledge for its study and all art for its expression. A painter, sculptor, engineer, architect, physicist, biologist and philosopher was Leonardo, and in each supreme. Perhaps no man in the history of the world shows such a record. His performance, extraordinary as it was, must be reckoned as small compared with the ground he opened up, the grasp of fundamental principles he displayed, and the insight with which he seized the true method of investigation in each branch of enquiry. If Petrarch was the harbinger of the literary Renaissance, in spirit Leonardo was in advance of it in other departments. He was not a scholastic, and neither was he a blind follower of classical authority, as were many of the men of the Renaissance. To him, observation of nature and experiment were the only true methods of science. Knowledge of the ancient writers, useful as a starting-point, could never be conclusive.

Leonardo approached science from the practical side, and it is owing to this lucky fact that much of his modern spirit is due. To meet the practical necessities of his crafts, he began experimenting, though in his later years the thirst for knowledge overcame the love of art. It was needful for an artist to understand the laws of optics and the structure of the eye, the details of human anatomy, and the flight of birds. As an engineer, both civil and military, Leonardo was faced by problems which could only satisfactorily be solved by an insight into the principles of mechanics, both static and dynamic. Now Aristotle's opinion was of small help in correcting a picture out of drawing, in managing water for irrigation, or in taking a fortified city. For these problems, the behaviour of things as they were was of more importance than the opinion of the encyclopædic Greek as to what they ought to be.

But Leonardo was also a philosopher, and as a philosopher he probably most affected the thought and mental attitude of his contemporaries. The most striking change we observe when comparing his mode of thought with that of the preceding age, is his almost complete emancipation from theological preconceptions. Even Roger Bacon, with all his love of enquiry, regarded theology as the true summit and end of all knowledge, and doubted not that all learning would prove consistent with the chief dogmas of his day.

But Leonardo reasons with a perfectly open mind. When he turns to theology at all, he attacks openly and lightly the ecclesiastical abuses and absurdities which had become part of the system of the Church. His own philosophical position seems to have been an idealistic pantheism, in the light of which he saw everywhere the living spirit of the Universe. Yet, with the fine balance of a great mind, he saw the good beneath the load of inconsequent evil, and accepted the essential Christian doctrine as an outward and visible form for his inward spiritual life. " I leave on one side the sacred writings," he says, " because they are the supreme truth." A great gentleman as well as a great man, the fanaticism of the rude iconoclast was far from Leonardo, and he lived in the brief interval when the Papacy itself was liberal and humanist, and all seemed pointing to a new and comprehensive Catholicism, in which freedom of thought could exist side by side with earnest mystical faith. The dream passed, the Church of Rome became reactionary, and freedom was won painfully and slowly by the rough path opened by Luther. Fifty years later, Leonardo's position would have been impossible.

Leonardo's note-books have enabled us to trace the origins of modern science in a way quite impossible before. Da Vinci, great as he was, must not be represented as the originator *de novo* of the scientific spirit he displays. Before him Alberti (1404–1472) had studied mathematics and made physical experiments. At Florence Leonardo met Paolo Toscanelli (d. 1482), an astronomer who had instigated the voyage of Columbus ; Amerigo Vespucci gave him a book on geometry ; he knew Luca Pacioli the mathematician, and was helped in his anatomical researches by Antonio della Torre. It is clear that, a century before the days of Galileo, a small circle of kindred spirits lived in Italy who were more interested in things than in books, in experimental enquiry than the opinions of Aristotle.

But there is a link with Greek thought behind these men too, a link with Archimedes. Archimedes' books had not yet been printed, and good manuscripts were rare. Leonardo notes the names of his friends and patrons who could procure him copies, and expresses admiration at the genius of the great Syracusan. Interest in Archimedes grew rapidly ; in 1543 the mathematician Tartaglia published a Latin translation of some of his works, and other editions followed, so that they were well known by the time of Galileo, who studied them carefully. In Archimedes, the man of science, and not in Aristotle, the encyclopædic philosopher, is to be sought the veritable Greek proto-type of modern physical science, of which Archimedes alone of the ancient writers who have survived possessed the true spirit.

Leonardo da Vinci perceived intuitively and used effectively the right experimental method a century before Francis Bacon philosophized about it inade-quately, and Galileo once more put it into practice. Leonardo wrote no treatise on method, but incident-ally his ideas about it clearly appear. He knew that mathematics, arithmetic and geometry, gave absolute certitude within their own realm ; they were concerned with ideal mental concepts (*e tutta mentale*) of universal validity. But true science, he held, began with observation ; if mathematical reasoning could then be applied, greater certitude might be reached, but " those sciences are vain and full of errors which are not born from experience, the mother of all certainty, and which do not end with one clear experiment (*che non terminano in nota experientia*)." Science gives certainty, and science gives power. Those who rely on practice without science are like sailors without rudder or compass.

When we turn from Leonardo's method to his actual results, we are astonished at his insight. In face of all the prepossession of the centuries, of the universal belief that all motion must be maintained by a cause continually acting, he enunciates the principle of inertia, afterwards demonstrated by Galileo. " Nothing perceptible by the senses," says da Vinci, " is able to move itself ; . . . every body has a weight in the direction of its movement." He knows that the speed of a falling body increases with the time, though he misses the right law for the space fallen through.

He clearly understands the experimental impossibility of " perpetual motion " as a source of useful power, and inveighs against those who attempt it. He uses this principle to demonstrate the law of the lever by the method of virtual velocities, hitherto attributed to Ubaldi and Galileo. The shorter arm raises the greater weight slowly, while the longer arm is pushed down by the smaller weight quickly ; these motions must be in the proportion of the lengths of the arms, so that the weights must be inversely as the lengths of the arms. Leonardo regards the lever as the elementary machine, and all other machines as modifications and complications of it.

Leonardo recovered Archimedes' conception of the pressure of fluids. He showed that liquids stand at the same level in communicating vessels, while, if different liquids fill the two arms, their heights will be inversely as their densities. He deals also with hydrodynamics—the efflux of water through orifices, its flow in channels, the propagation of waves over its surface. From waves on water he passes to waves in air and the laws of sound, while he saw that light showed many analogies which suggested that here too a wave theory was applicable. The reflection of an image is the echo of the source ; as with a ball thrown against a wall, the angle of reflection is equal to the angle of incidence.

In the realm of astronomy Leonardo conceived of a celestial machine conforming to definite laws, in itself a remarkable advance on the prevalent Aristotelian ideas that the heavenly bodies are divine, incorruptible, essentially different from our world subject to change and decay. He calls the earth a star, not different from the others, and proposes in his projected book to show that it would reflect light like the moon. With errors in detail, Leonardo's astronomy is true in spirit, and with him modern astronomy appears.

As things are older than writings, the earth bears trace of its history before the records of books. Fossils now on high inland mountains were produced in sea-water, and could not have reached their present position in the forty days of the Noachian deluge ; indeed the whole waters of the world, clouds, rivers and ocean, could not cover the mountains of the earth.

There must have been changes in the crust of the earth, and mountains have raised themselves in new places. But no catastrophic action is needed ; " in time the Po will lay dry land in the Adriatic in the same way as it has already deposited a great part of Lombardy "—Hutton's uniformitarian theory three hundred years before its time !

As a painter and sculptor, da Vinci felt the need of an accurate knowledge of anatomy. In the face of ecclesiastical tradition, he procured many bodies and dissected them, making anatomical drawings which are works of art as well as most accurate in all detail. Many of them still exist in his manuscripts preserved at Windsor. " And you who say that it would be better to look at an anatomical demonstration than to see these drawings," he observes, " you would be right, if it were possible to observe all the details shown in these drawings in a single figure, in which, with all your ability, you will not see nor acquire a knowledge of more than some few veins, while, in order to obtain an exact and complete knowledge of these, I have dissected more than ten human bodies."

From anatomy the next step is physiology, and here too Leonardo is found in possession of the field. He describes how the blood makes and remakes continually the whole body of man, bringing material to the parts and carrying off the waste products, as a candle or furnace is fed. He studies the muscles of the heart, and makes drawings of the valves which in the opinion of Knox show a knowledge of their functions. He compares the flow of blood with the circulation of water from the hills to the rivers, from the rivers to the sea, from the sea to the clouds and back to the hills as rain. It seems that Leonardo understood the circulation of the blood a hundred years or more before it was rediscovered and Harvey gave the knowledge to the world. Once more his art led him to a scientific problem, in the structure and mode of action of the eye. He made a model of its optical parts, and showed how an image was formed on the retina. He ignores the view, still held by his contemporaries, that the eye throws out rays which touch the object it wishes to examine.

He dismisses scornfully the follies of alchemy, astrology and necromancy; for him nature is orderly, non-magical, subject to immutable necessity.

But we have said enough to illustrate Leonardo da Vinci's position in the history of scientific thought. Had he published his work, science must have advanced by one step to the place it reached at least a century later. It is idle to speculate on the influence of such a change on the story of the human mind, and the evolution of human society. It is safe to say that both would have been modified profoundly.

But, though Leonardo never carried out his oft referred to project of writing books on different branches of his labours, his personal influence was clearly immense. The friend of princes, he knew also all the men of learning and affairs of his time. His ideas were not all sterile, but some fell on the good ground which, years later, gave birth to a new growth of scientific achievement, springing from the seed scattered by Leonardo and his disciples. If we had to choose one figure to stand for all time as the incarnation of the true spirit of the Renaissance, we should point to the majestic form of Leonardo da Vinci.

8

In a society stirring with so many intellectual interests we have a mental environment very different from that of a hundred years earlier. The theological atmosphere, which saw everything in the light of the one overpowering motive of salvation, had given place to a much more independent outlook, in which many questions were freely discussed from a rational point of view. The world was still orthodox, but orthodoxy itself had been aroused and for a time stretched its bounds.

Copernicus.

Nicolaus Koppernigk (1473–1543), a Polish mathematician and astronomer, had long been dissatisfied with the prevalent Ptolemaic system, and he returned from a long stay in Italy, where the Pythagorean heliocentric theory was now well known as a Greek speculation, determined to put it to the proof. With scanty instrumental resources he made a series of observations, and worked out his theory. He showed how much simpler it was as an explanation of the phenomena than the Ptolemaic system of cycles and epicycles, in which the heavenly bodies moved round the earth as centre. He finished a treatise setting forth his scheme about 1530, and published a short abstract in popular form in that year. Pope Clement VII. approved, and sent the author a request for the publication of the work in full. To this Copernicus only consented in 1540, and the first printed copy reached him on his deathbed in 1543.

But if in 1530 the Papacy showed a liberal interest in the new system, by 1616 it had determined, by the mouth of Cardinal Bellarmine, that it was " false and altogether opposed to Holy Scripture," and Copernicus' book was suspended till corrected. As no one cared to issue an edition with the authorized corrections, it remained prohibited for over two hundred years, for it was not till 1822 that the sun received the formal sanction of the Papal authorities to become the centre of the planetary system.

In the year 1822 the Roman Church at last removed the works of Copernicus from the Index of prohibited books and permitted instruction in the heliocentric theory. Nevertheless, the Papal bulls by which it is forbidden to believe in the motion of the earth still remained in force, thus creating an extraordinary predicament in the scientific beliefs of the devout Roman Catholic.

It is very unfortunate for the historian of the progress of science and of the human mind, that the Roman Index is far from containing a complete catalogue of works deemed at various periods to be inconsistent with Catholic orthodoxy. It has thus lost a great deal of the importance and interest that otherwise it might have possessed, for the student of the development of thought.

The system of Copernicus taught men to look on the world in a new light. Instead of floating at the centre of the Universe, the earth sank to the lowlier place of one among the planets. Such a change does not necessarily involve the dethronement of man from his proud position as the summit of creation, but it certainly suggests doubts of that belief.

Bruno.

Thus, besides destroying that Ptolemaic system which had been incorporated as a necessary part of his scheme by Thomas Aquinas, Copernican astro-

nomy affected the human mind in other ways.   Certainly it was one of the influences which led the fiery
Dominican Giordano Bruno (*c.* 1548–1600) to break
with the Roman Church, and to become an outcast
both from Catholic Rome and Calvinist Geneva.
Bruno used the work of Copernicus to discredit the
authority of Aristotle, for whom he had a perfect
hatred, contrasting him most unfavourably with the
older Greek philosophers.   Passing from Paris to
England and England to Germany, Bruno everywhere attacked openly the Roman Church, its clergy,
and the doctrines they taught : the Jewish scriptures
he treated as myths, the miracles of the saints as
magical tricks.

Accepting a treacherous invitation to Venice in
1596, Bruno fell into the clutches of the Inquisition.
He was imprisoned, sent to Rome in 1593 and burnt
at the stake in 1600.

At last we see the ideas concealed in the manuscripts
of Leonardo da Vinci, then doubtless floating about in
the minds of his Italian successors, coming
to the light of day in the epoch-making
work of Galileo Galilei (1564–1642).   In Leonardo, the
spirit of modern science is present everywhere, framing, moulding and developing his thoughts on all the
innumerable subjects on which he pondered.   But in
Galileo it has gone further.   With an equally sound
grasp of principle, he has learnt the modern need of
concentration, and works out his more limited problems
in a more complete and methodical way than the
wider sweep of Leonardo's soul could stop to accomplish.   In reading Galileo we feel at once that modern

science is not only born, but has come to an age of steady work. Moreover, he collected and published his researches; and thus gave them at once to the world.

As an astronomer, Galileo's fame rests on his application of the telescope and the confirmatory evidence he thus gave to the Copernican system. It was the expression of his conclusions from these researches, embodied in a volume of which the dedication was originally accepted by the Pope, that brought Galileo, when his book had been read and understood, within reach of the Inquisition, and forced him to recant his heresies.

But Galileo's chief and most permanent work was the foundation of the modern experimental and mathematical science of mechanics. In this he showed himself a true master of method.

Before Galileo, in spite of Leonardo's superhuman insight, most men's ideas of mechanics were a confused medley of Aristotelian dogmas. Bodies were thought to be intrinsically heavy or light and to fall or rise with varying velocity because they " sought their natural place " with varying power. Galileo set himself to discover not *why*, but *how*, things fell. In itself this attitude of mind marks a great achievement.

A body falling towards the ground moves with constantly increasing speed. What is the law of the increase ? Galileo's first hypothesis, quite reasonable in itself, was that the speed was proportional to the distance fallen through. But Galileo soon showed that this supposition involved a contradiction, and tried another, namely that the speed increased with the time of fall. This hypothesis was found to involve no obvious difficulty, and Galileo proceeded to deduce its consequences, and to compare them with the results of experiment.

A body falling freely moves too fast for easy and accurate observation on its rate of fall, and, to bring the speed within convenient limits, Galileo convinced himself that a body falling down an inclined plane acquired the same velocity as though it had fallen freely through the same vertical height. He then experimented with inclined planes, and found that the results of his measurements agreed with those calculated from his hypothesis.

But another result, equally important, which had already been grasped by Leonardo, followed from Galileo's investigations with inclined planes. He found that, after running down one plane, a ball will run up another to a height equal to that of its starting-point, whatever be the slope, provided that friction be negligible. The second plane may be made as long as we please, but still, if the final height be the same as the initial height, the ball will reach its end. It is the height that matters ; the speed of the ball is acquired by virtue of its descent and is not destroyed unless the ball rise. And, if the second plane be horizontal, the ball will run along it for ever with uniform velocity, until checked by friction or some other force.

To appreciate the importance of this result, it is necessary to realise that, before Galileo's day, except by Leonardo and those he influenced, it was assumed that every motion required the continual exertion of some force to maintain it. The planets had to be kept in motion by hypothetical vortices moving through the heaven and carrying them round in their orbits, and similar complications appeared on every side in mechanical problems. By Galileo's investigation, the whole position was reversed, and it was seen that it was the destruction of motion or a change in its direction which required the exertion of some external applied force. Thus the planetary system needed no vortices to keep it in motion, but some cause was required to explain the continual deviation of the planets from a straight course, as they swung round the sun in their orbits. Never before had it been possible even to formulate the problem correctly, but now the way was open to a solution, and the man was at hand. In 1642, the year of Galileo's death, Isaac Newton was born.

But, before we follow the consequences of Galileo's labours, and deal with Newton's supreme achievement, Gilbert of Colchester. we must pass in review other trains of thought which went to make that new intellectual world in which Newton dwelt.

In the land of Newton the new method of experiment, so well used by Galileo in Italy, was simultaneously and worthily put in action by William Gilbert of Colchester (1540–1603), fellow of St John's College in Cambridge, and sometime President of the College of Physicians. In his book, *De Magnete*, Gilbert collected all that was known about magnetism, and added many new and valuable observations of his own.

The mariner's compass is said to have been known to the Chinese from early times. Descriptions of it seem first to have appeared in European literature about the twelfth century, coming probably from Saracenic sources, though the references indicate that

it had been used for some time.   Gilbert pointed out
that the set of the compass with reference to the earth
indicated that the earth itself possessed magnetic
properties, and could be represented as a huge magnet,
with magnetic poles nearly but not quite coincident
with the geographical poles.

To Gilbert also we owe the name electricity, derived
from the Greek word ἤλεκτρον, amber, a substance
which becomes electrified when rubbed.   Gilbert
investigated the forces due to such electrification by
using a light metallic needle balanced on a point, and
extended the number of bodies which showed the
effect.

It is worthy of note, as an indication of the appreciation of such researches by the Government of England,
that Gilbert, who was Court physician to Elizabeth
and James I., was given a pension to set him free to
continue his experiments in physics and chemistry.

Meanwhile the philosophy of the new experimental
method was considered by Francis Bacon (1561–1626),
Lord Chancellor of England.   Bacon was
Francis Bacon. deeply impressed by the failure of the
scholastic philosophy to advance man's knowledge of
and power over nature.   He felt that mankind was
still but the plaything of untamed forces, and had not
yet grasped that *imperium hominis* which should be
his birthright.   " To extend more widely the limits
of the power and greatness of man," he laid down rules
by which advance towards mastery might be made
more sure.

Bacon taught that by recording all available facts,
making all possible observations, performing all

feasible experiments, and then by collecting and tabulating the results, it would be possible to determine what phenomena varied together, and thus to discover the true and inevitable relations between them.

The obvious criticism of Bacon's method is that, while partially applicable to purely descriptive sciences like natural history, it is never applicable elsewhere. The numbers of phenomena and possible experiments are too numerous to be treated thus. At an early stage of the enquiry, scientific insight and imagination must come into play to exercise a selective action ; a tentative hypothesis must be framed, and the multitude of possible experiments reduced to the manageable number needed to confirm or refute the hypothesis. Hypothesis plays an essential part in science, and research seldom or never proceeds on pure Baconian lines.

Many years later, the method of Bacon was criticized by T. H. Huxley in the following words :—
" Those who refuse to go beyond fact," he wrote, " rarely get as far as fact ; and any one who has studied the history of science knows that almost every great step therein has been made by the ' anticipation of nature,' that is, by the invention of hypotheses, which, though verifiable, often had very little foundation to start with ; and, not unfrequently, in spite of a long career of usefulness, turned out to be wholly erroneous in the long run."

There is much that compels our assent in this nineteenth-century criticism of the Baconian method ; but we must remember that up to the time in which Bacon wrote, the world had listened to many theories and hypotheses, and had seen no corresponding accumulation of facts whereby to test them.   Rightly therefore, in Bacon's eyes, facts—observed, authenticated facts—were the crying need of his age.

Although Francis Bacon himself made no contributions to natural knowledge, and although his treatment of method was over-ambitious in range and incapable of general application in practice, he was nevertheless a foremost figure in the early advance of modern science.   In terms of conscious power and statesmanlike eloquence, he expressed ideas that were floating inarticulately in the society around him. The authority of theology in the realm of natural philosophy had been set on one side, at any rate in England.   The doctrines of the schoolmen had been both outgrown and worn out.   The world of thought was astir, and time was ripe for a change.   Bacon gathered up the scattered threads, set the ship of progress firmly on what was roughly the right track, sped it on its way and gave to it its sailing directions in the *Novum Organum*.

Some of the work that was accomplished by Francis Bacon in England, was carried out somewhat earlier in France, but in a very different way, by Michel de Montaigne.   Montaigne (1533–1592) was born in the family château, not far from Bordeaux, in a district where much Northern blood had been left behind after the English occupation ;   and, like his father, served in the army and in the local parliament.   Living at a time when the French Renaissance was at its height, when, indeed, the tide was beginning to turn, Montaigne's essays foreshadow the coming period of disenchantment and give

The Renaissance in France.

voice to the change of attitude which was then mani-
festing itself throughout the country. All aspects
of human life and activity, all varieties of belief
interested him profoundly, and are set down in his
famous volumes with an impartiality of outlook
which gives a strong impression of scepticism and
independence. The *Essays* were widely read and
translated, and exercised an enormous influence,
more than their contents would seem to justify,
showing that they were felt to be in harmony with a
frame of mind that must have been widely prevalent.

A generation later than Montaigne and Francis
Bacon, a man of French birth and ancestry laid the
foundations of modern critical philosophy.
Descartes. René Descartes (1596–1650) was born in
Touraine, of a family of the demi-noblesse, and died at
Stockholm in the service of Queen Christina of Sweden.
His chief work was accomplished during a twenty
years' sojourn in Holland, to which country he origin-
ally went to serve under Prince Maurice of Orange
in the wars of independence against Spain.

The distinguishing feature of Descartes' work was
its combination of far-reaching and sound mathematical
advance, somewhat speculative physical theory and
a critical philosophy which took nothing for granted.
Descartes showed how much unverified assumption
underlay even generally received philosophic ideas,
and from him modern " philosophic doubt " took its
origin : *de omnibus dubitandum est.*

Descartes turned from the old accumulations
of interwoven thought, and tried to build up a new
philosophy, based only on human experience and

consciousness. As he found in human consciousness
an apprehension of God, he passed neither to materialism nor to that complete divorce between faith and
reason which, necessary for the well-being of both
at a certain stage of development, leads eventually to
the decay of faith.   Descartes indeed remained a good
Catholic.   But his belief in the possibility of proving
the existence of God by philosophic evidence savoured
of heresy to the theologians both of Utrecht and of
Leyden, and, more than once, Descartes had to appeal
to the protection of the Prince of Orange.

In mathematics, Descartes' greatest achievement was
the application of algebra to the problems of geometry,
in which hitherto each problem had to be solved by
special treatment and by a fresh display of ingenuity.
For the first time, Descartes introduced a general
method by which this isolation was broken down.
The primary idea of co-ordinate geometry is easily
stated.   If two straight lines or axes start from a
point or origin and set out at right angles to each
other, it is possible to specify the position of any point
in their plane by stating its distance x from one axis
and its distance y from the other ; x and y are called
the co-ordinates of the point, and different relations
between x and y correspond to different lines or curves
in the plane of the diagram.   Thus if y be proportional
to x, or $y = x \times$ constant, we get a straight line; if
$y = x^2 \times$ constant, a parabola; and so on.   The problems of geometry are reduced to the operations of
algebra.

Descartes' physical speculations were less successful
than his mathematical discoveries.   We hear an echo
of the old ideas of contrasted words which deceived

many a Greek philosopher. Descartes contrasts matter and spirit, and, since spirit is individual and exists in personal units, matter must be continuous. In a continuous, closely packed universe, movement is only possible if it occur in closed circuits, every part of which must move together. Hence Descartes arrived at his famous theory of vortices, which for twenty years reigned supreme, and contested for a time even the Newtonian system, which the Cartesians argued gave no explanation, since it depended on the hypothesis of unknown and mysterious forces.

Yet Descartes' physical ideas did good service to science. Though barred by Protestant orthodoxy, and put on the Catholic Index at the instigation of the Jesuits, his works became the fashion. They offered an explanation of the phenomena of the astronomical universe by mechanical processes ; they banished for ever Aristotle's distinction between the essential nature of the sublunary sphere, which included the earth, and the starry sphere beyond of incorruptible and perfect heavenly bodies ; they made unnecessary the animate being of Aquinas or the genii believed in by Kepler as the cause of planetary motion. Whether or no the Cartesian explanation stood, the solar system was susceptible of physical treatment ; in common words, the thing could be understood.

Descartes spent most of his working lifetime outside France, and seems to have been to a great extent independent of contemporary French influences, such as that exercised by Montaigne—or, at any rate, unconscious of them ; but in another direction the sceptical attitude of current

Pascal.

thought came directly within the ken of natural science. Blaise Pascal (1623–1662) always acknowledged his indebtedness to the writer of the famous *Essays*; and the whole of the Jansenist colony of Port-Royal, with their rationalist and mystical outlook, in which the Pascal family were deeply involved, displayed much of Montaigne's attitude of mind in their dealings with the orthodox Roman Catholic Church. Pascal was born at Clermont-Ferrand, in the Auvergne, the family having been ennobled during the fifteenth century. As a philosopher, he is best known by his *Lettres à un Provincial*, attacking the teaching and methods of the Jesuits; as a mathematician he worked, like Descartes, to generalize the theory of endless particular propositions, and he was the founder of the mathematical theory of probability, a study which originated in a discussion concerning the division of stakes in games of chance. His experiments and subsequent treatise on the equilibrium of fluids place him with Stevin and Galileo in developing the science of hydrodynamics; while his direction of the famous experiment with the barometer on the Puy de Dome, which showed that the height of the mercury column did indeed diminish as the instrument was carried upward, brought home to men's minds the meaning of the discoveries of Galileo and Torricelli.

But if Galileo and Descartes cleared and prepared the ground for Newton, the subject-matter which

Tycho Brahe and Kepler. his genius moulded into shape had been put forth ready to his hand, chiefly by Tycho Brahe and John Kepler (1571–1630).

The planetary motions had been measured of old

and described in terms of Ptolemaic cycles and epicycles. But, inspired by the Copernican ideas, Brahe, working on his island home off Copenhagen, carried such observations to a higher degree of accuracy than had ever before been known, and the accumulations of his chequered lifetime fell into the hands of an apt and eager follower.

Working with Brahe's results, Kepler, whose official occupation consisted chiefly in calculating the astrological almanacks which, at that period, were in favour at the petty courts of Germany and Austria, after most laborious investigations, found a series of laws which described the motion of the planets. He showed, for instance, that they travelled in paths which were ellipses, and that the sun was at one of the two foci of the orbits, and this result, with his other laws, led Newton to apply his unrivalled mathematical powers to the planetary problem.

But we must pause yet once more to trace another influence in Newton's intellectual environment. The Scientific number of those interested in natural Academies. philosophy was now increasing rapidly, and one sign of this increase was the establishment of societies or academies consisting of men who met together to discuss the new subjects and to further their interests. The earliest of such societies appeared at Naples in 1560 under the name of Academia Secretorum Naturæ. In 1651 Florence followed by founding the Accademia del Cimento. In England a society began in 1645 to meet at Gresham College or elsewhere in London. In 1648 most of its members moved to Oxford owing to the Civil War, but in 1660 the meetings in London were revived, and in 1662 the society was formally incorporated by charter of Charles II. as the Royal Society of London for Promoting Natural Knowledge. In France the corresponding Académie des Sciences was founded by Louis XIV. in 1666, and similar institutions soon appeared in other countries. Their influence in focussing scientific opinion, and making known the researches of their members, has had much to do with the more rapid growth of science since their foundation.

Isaac Newton (1642–1727), by universal consent the greatest man of science of all time, the delicate,

Newton. posthumous and only child of a yeoman, was born at Woolsthorpe in Lincolnshire, and educated at Grantham Grammar School and at Trinity College, Cambridge, where he was greatly influenced by the work of Descartes. About 1666, driven to Woolsthorpe by an outbreak of plague at Cambridge, he turned his attention to planetary problems. Galileo's researches had shown the need of a force to keep the planets in their orbits and prevent them moving off in a straight line. The Dutch physicist Huygens had calculated the intensity of the force needed to keep a body whirling in a circular orbit with a known velocity. It remained to show that such a force existed in the case of the moon and the planets.

Newton is said to have grasped the clue while idly watching the fall of an apple in the orchard at Woolsthorpe. If the earth pulled the apple, would it not also pull the moon ; and would not the sun by a similar force pull the planets round in their orbits ? The distance of the moon is 60 radii of the earth, so that the force would be less at the distance of the moon than at the surface of the earth in the ratio of $(60)^2$ or 3600 to 1. With the only estimate of the earth's radius available in 1666, this calculation gives a force too great to explain the moon's motion round the earth, and Newton, always dissatisfied with the slightest inaccuracy, put aside the enquiry. But in 1679 a redetermination of the size of the earth, giving a markedly different value, led Newton once more to take up the problem, and, in a state of excitement which is said to have been so great that he could hardly see his figures, he proved that the fall of a stone to the earth and the majestic sweep of the moon in her orbit may be ascribed to one and same cause.

Newton then proceeded to attack the general problem of the motion of a body under a force directed towards a fixed point, and showed that the supposition that every particle of matter attracts every other particle in accordance with the law of inverse squares was necessary and sufficient to explain Kepler's planetary observations. The simple law of the heavens, the universal rule of gravity, was thus revealed—the greatest achievement in the history of science.

The mechanism by which the force is exerted remained unexplained by Newton, and has hitherto baffled all subsequent enquiry. As always, one of the chief results of new scientific knowledge was to define more accurately the limits of our ignorance. Newton indeed described himself as a child gathering pebbles of knowledge by the shore of the boundless ocean of the unknown. The larger grows the sphere of knowledge, the greater is its surface of contact with what lies beyond, and the clearer are the gaps within it.

One of the immediate results of the application of mathematical mechanics to the problems of astronomy was the need of improvement in the mathematical tools used in the researches. Hence the same period which saw the labours of Kepler, Galileo, Huygens and Newton was marked also by a great increase in mathematical knowledge and skill.

Perhaps the most noteworthy of these achievements was the invention of the infinitesimal calculus, developed by Newton and Leibniz. Algebra and geometry had begun to assume their modern shapes, trigonometry had been extended to imaginary quantities, but the introduction of the idea of varying velocity demanded a method of dealing with the rates of variation of changing quantities. A constant velocity is measured by the space s described in a time t, and the quantity s/t will be the same however great or small s and t may be. But, if the velocity vary, its value at any instant can only be found by taking a time so short that the velocity does not change appreciably, and measuring the space described in that short time. When s and t are reduced without limit and become infinitesimal, their quotient gives the velocity at the instant, and was written by Leibniz as ds/dt, which is called the differential coefficient of s with regard to t. Newton, in his method of fluxions, wrote the same quantity as $\dot{s}$, a notation which is less convenient and is now superseded by that of Leibniz. We have taken space and time as an example, but any two quantities which depend on each other may be treated in the same way, and the rate of variation of x with y written as dx/dy or $\dot{x}$. These conceptions may be said to be the starting-point of the whole vast structure of modern pure and applied mathematics.

Newton also took the lead in the development of physical optics. In 1669, when he was appointed to a professorship at Cambridge, he chose optics as the first subject of his lectures and researches. The discovery of the decomposition of white light by the prism, and the explanation of the rainbow, soon followed, as did the invention of a new form of reflecting telescope. Two theories of light were known. On the corpuscular or emission theory, luminous bodies emit a stream of minute particles or " corpuscles " which by impact affect the eye. On the other theory, light consists of waves or undulations, and on this view it appeared necessary to revive a conception which we owe originally to the Greek philosophers; and thus it has been usual, till lately, to invent a subtle medium or æther, which is imagined to possess certain material properties and to fill all space. Owing to the difficulty of explaining the straight path of light, Newton inclined to the corpuscular theory, while Huygens did much to develop the theory of waves. It is certain that the great weight of Newton's name helped to prevent the earlier acceptance of Huygens' views—another instance of the danger of too much reliance on authority. But the recent undoubted discovery of the existence of moving particles, closely resembling Newtonian corpuscles, and shot off from radio-active bodies with velocities approaching that of light, has once more made clear the almost superhuman insight into nature possessed by Newton's mind.

We must now return to the influence of the Renaissance on other branches of science, and, in chief, on

Medical Science during the Renaissance.

medicine, from which most of the others took their rise. It was at first supposed that the revival of Greek learning would produce the same brilliant results in medicine as in literature and philosophy ; and a school of medical humanists arose who from 1450 to 1550 turned men's minds from mediæval medicine, developed through Arabian channels from commentaries on Greek writers, to what were regarded as the fountain-heads of the science—the writings of Hippocrates and Galen themselves.

But even from the first there were revolts against the domination of the newly translated authorities, and, once again, an actual observation of nature formed the starting-point. Paracelsus (c. 1490–1541), a man of Swiss birth, broke away from the schools and the chemistry of his period, and in the mines of the Tyrol studied indifferently rocks, minerals, mechanical contrivances and the conditions, accidents and diseases consequent on the miner's life and occupation. He then wandered over a great part of Europe studying the diseases and remedies of different nations, before settling down for a while as " town physician " at Basle, where he roused the opposition of the vested interests by the efficacy of his treatment and the guaranteed purity of his drugs. With independent arrogance, Paracelsus taught in the German language contempt for Galen and Avicenna, whose works he burnt publicly in the lecture-room, and relied on his own experience, interpreted in the light of a personal reading of neo-Platonic philosophy. But perhaps his most lasting innovation was the application of the chemical knowledge, gained by alchemists in their search for gold, to medical problems. "Chemical medicine" marked the follower of Paracelsus, and for long distinguished him from an orthodox Galenic school. Chemistry began to develop in new directions when it was studied for the sake of discovering substances to cure disease, as well as for the illusive vision of metallic transmutation.

In 1543, Andreas Vesalius, a Fleming by birth, trained in the French school and afterwards professor simultaneously at Padua, Bologna and Pisa, published a book on human anatomy, founded, not on what Galen taught, but on what he himself had seen in dissection and was prepared to demonstrate in the lecture-room. But he was denounced to the Inquisition, forced as a penance to undertake a pilgrimage to the Holy Land, was wrecked and probably devoured by wild beasts on the coast of Zante.

Nevertheless, before the end of the sixteenth century, anatomy, first of all the sciences, was freed from the William   trammels of ancient authority. Physiology Harvey.   lay longer in bondage, till William Harvey, who had studied in Italy, was led " to give his mind to vivisections," and thereby made clear the true mechanism of the circulation of the blood—" of motion as it were in a circle."

William Harvey's life and work deserve more than a passing reference. He was born in 1578, the son of a prosperous Kentish yeoman or small squire, and died in 1657. After studying in Cambridge, he spent five years abroad, chiefly in Padua. Returning to

England when he was about twenty-four years of age, he began to practise as a physician, numbering Francis Bacon himself among his patients. He was in attendance on James I., and it fell to the lot of the most modern physiologist of the time to superintend the medical examination of women accused of witchcraft. With Charles I. Harvey was on terms of intimacy. The King had placed the resources of the deer parks at Windsor and Hampton Court at the disposal of his Court physician for experimental purposes, and, with him, watched the development of the chick in the egg and the pulsations of the living heart. Harvey followed the King and was in charge of the royal princes at the battle of Edgehill, viewing the fight from the rear. He retired to Oxford with his master, where for some time he was Warden of Merton College. Having no children, he bequeathed his paternal estate to the Royal College of Physicians, directing them to use the proceeds " to search out and study the secrets of nature."

The essence of Harvey's great discovery lay not so much in demonstrating the circulation of the blood in the veins, which indeed had already been made clear by the anatomists of the sixteenth century, as in revealing the mechanism by which the circulation was maintained. The existence and use of the valves in the veins was also known, and Michael Servetus (1511–1553), the Aragonese physician and theologian who was burnt at Geneva by Calvin's orders, had possessed himself of many of the actual facts of circulation. Harvey's claim to fame lies in his correct setting forth of the true meaning of the action of the heart, as the organ responsible for maintaining the whole circulation, and in his convincing correlation of all parts of the circulatory system. He described accurately everything that was visible to the naked eye, so that it remained only for Malpighi, a few years later, to point out, in work communicated to the Royal Society of London in 1672, the capillary tubes by which the blood passes from the arteries to the veins, at that time first made visible by means of the newly invented microscope.

Harvey's first treatise, a small volume entitled, *Exercitatio Anatomica de Motu Cordis et Sanguinis*, was printed at Frankfort in 1628 and dedicated to Charles I. His second and larger work, *De Generatione Animalium*, was published in 1651, and contains almost the first advance in embryology recorded since the time of Aristotle.

Throughout the sixteenth and seventeenth centuries, new chemical substances were quickly coming to The Rise of light, discovered in the search for fresh Chemistry. remedies and industrial materials. But, for some time, there was no corresponding advance in chemical theory. Chemists continued to accept the ancient view of three " elements " or " principles " in the form of sulphur, mercury and salt. At last, in 1661, Robert Boyle, a younger son of the great Earl of Cork, in his *Skeptical Chemist* attacked the prevailing views and revived the atomic theory, incidentally indicating the true nature of heat. But for this step in advance the time was not yet ripe, and the conceptions which proved best suited to the needs of the age, and by which the immediate advance in knowledge was won, were much less in accordance with quite modern views.

George Ernest Stahl (1660–1734), physician to the King of Prussia, developed the idea of a volatile "principle" to explain the phenomena of combustion. When bodies were burnt something apparently escaped, and this something Stahl named phlogiston, the "principle of fire." Although this view regarded burning as a loss of substance, and was thus in contradiction to facts known to Boyle. who had shown that metals increased in weight on burning, it was so powerfully advocated, and expressed so well the ideas of the age, that it gained general acceptance, and dominated the chemical ideas of the whole eighteenth century. Stahl also opposed the somewhat crude, and at all events premature, materialistic views which were beginning once more to appear in physiology. He held that the chemical changes in living bodies, though carried on in accordance with those produced in the laboratory, were directed and governed by the "sensitive soul" described by Aristotle—a view which lived long under the name of vital force, and in another form is perhaps once more coming to be thought necessary to explain the phenomena of the living organism as a complex adaptive synthesis of matter and energy.

During the seventeenth and eighteenth centuries the systematic exploration of the world began to take Voyages of its place in the organised pursuit of know-
Discovery. ledge. If the explorers of this period cannot claim for their voyages the romance associated with the pioneers of discovery in the fifteenth and sixteenth centuries, pioneers who first revealed the existence of the earth as we now know it and mapped

the main configuration of continents and oceans, the work of the later navigators is remarkable for the growth of the scientific spirit of observation.

William Dampier (1653–1715) in particular showed the new attitude of mind. His keen eye noted every strange bird and beast, every new tree and plant, and his facile pen described their forms and hue with marvellous accuracy and recorded them in his volumes of *Voyages*. His *Discourse on Winds* became a classic of meteorology, and Humboldt, long afterwards, praised him as the best hydrographer of his age.

The development of medicine in the treatment of disease by drugs soon reacted on the knowledge of plants,
Botany and Natural History.
originally a province of the traditional lore of monastery and convent garden. Mediæval symbolism was slow to loose its grip of the plant world, where it took the form of the doctrine of " signatures," and regarded the shape of the leaf or the colour of a flower as an index or sign of the use for which the plant had been intended by its Creator. However, the increased security of life led to the laying out of private gardens and parks, and to the more general cultivation of trees, vegetables and flowers. Thus, partly owing to the use of herbs as remedies, partly to natural curiosity and to a growing love of beauty and colour, made possible by the advances in ordered existence, the sixteenth century saw a great development in botanical knowledge. Medicine, virtually freed from the control of the Church, soon acquired its own gathering grounds and distilleries. Botanic gardens were established at Padua in 1545, and afterwards at Pisa, Leyden and

elsewhere, and there the rare plants brought home by the explorers and adventurers were deposited and cared for. Each society of apothecaries had its physic garden, one of which yet survives amid the crowded thoroughfares of London. A number of " herbals " containing descriptions of plants and their properties—medicinal and culinary—began to appear and to find their place in the libraries of the country gentlemen. Such an one, illustrated by woodcuts, was published in England by John Gerard in 1597, Gerard himself being a member of the court of Barber-Surgeons and superintendent of Lord Burghley's new gardens at his house near Stamford.

The anatomy of plants was studied as soon as the microscope made the subject possible for investigation, and, led by Malpighi and Nehemiah Grew, men began to form correct ideas about the functions of the different plant organs. Although from early times the fruit of plants had been recognized to be connected with a female element, it was not till the end of the seventeenth century that definite experimental proof was given by Camerarius (1665–1721), to show that the anthers were the necessary male organs, that in their absence no fertilization of the female element was possible, and that without fertilization seed could not be formed.

The earlier classifications of both animals and plants were chiefly based on utilitarian ideas—or on such obvious external signs as led to the division of plants into herbs, trees and shrubs. Ray, by giving definiteness to the idea of species, pointed the road to a more natural scheme.

It was on the sex-organs of plants that Linnæus

(1707–1778), the botanist, son of a Swedish clergyman, founded his famous system of classification. Linnæus' classification held its own till replaced by the modern system, which, by considering all the characters of the organism, tries to place plants in groups which express their natural relationships.

Linnæus also turned his attention to the varieties of the human species, having been struck by the obvious differences of race brought to his notice during his wanderings among the Laplanders in search of arctic plants. In his *System of Nature* he placed man with apes, lemurs and bats in the order of " Primates," and subdivided man into four groups according to skin colour and other characteristics.

Development in the knowledge of animals was stimulated by the arrival in Europe of rare and strange beasts to grace the various royal menageries. The comparison of different organs, and the study of the functions of different parts of the bodies of man and of other animals, led to a growth of comprehensiveness of outlook, while improvements in the microscope allowed a far closer insight into the problems of the structure and function of the different organs of the body, and soon disclosed the existence of vast classes of minute organisms unsuspected before.

Buffon (1707–1788), the first of the great French naturalists, was born in Burgundy of a good family. After travelling on the Continent, he settled in Paris, where he was appointed keeper of the Royal Gardens and Museum. Buffon's genius, as well as his opportunity, lay in descriptive work, and his *Natural History of Animals* has all the merits of an encyclopædia. While not consenting with Linnæus to classify man among the animals — " une vérité humiliante pour l'homme "—Buffon could not close his eyes to the evidence pointing towards such relationships, and ventured the remark, which he was afterwards obliged to withdraw, that had it not been for the express statements of the Bible, one might be tempted to seek a common origin for the horse and the ass, the man and the monkey.

Both in ancient times and in the Middle Ages, men firmly believed that living things might arise Spontaneous *de novo* from dead matter. "From Aris-Generation. totle to Augustine, from Lucretius to Luther, the belief in spontaneous generation remained unshaken." Frogs might be generated from mud by sunshine, and perhaps the aboriginal Americans, whose descent from Adam was difficult to trace, might have the same origin. But Francesco Redi (1626–1697) showed that, if the flesh of a dead animal be protected from insects, no grubs or maggots appeared in it.

Redi's experiments were considered to controvert the Scriptures, and were attacked on that ground ; an interesting fact in the light of its reversal in recent years, when an unsuccessful attempt to prove spontaneous generation was prematurely reprobated as going to show that life might arise without direct creation. It seems as though the theological mind when ill-regulated hates a novelty for its own sake. But it is refreshing to find that Redi's work was upheld and extended by the Abbé Spallanzani (1729–1799), who showed that even minute forms of life did not develop in decoctions which had been boiled vigorously and then protected from the air.

We cannot close an account of the progress of scientific thought during the Renaissance and the succeeding years without some reference to witchcraft and to the belief in the active interference of unseen and generally speaking evil forces in the affairs of mankind. At the time when Bacon was directing the advance of knowledge in England, when Luther, Calvin and Knox were reforming the churches of Germany, France and Scotland, a belief in witchcraft dominated practically the whole civilised world. It is easy now to laugh at such fears, to ridicule the stories which were then implicitly believed, and to explain away the long series of occurrences on which the epidemic of prosecutions throughout Western Europe was based. But we must face the fact that the movement went on side by side with the Renaissance, and to some extent seems to have been the popular representation of a belief in that possibility of man's control over the unknown forces of nature which, on the scientific side, Bacon foresaw and desired. The particular form which the accusation usually took, that of commerce with the powers of darkness for evil purposes, was probably due to the prominence given to the personality of the Devil in the Protestant theology of the day. Many of the great men of science of this period were either accused of heresy or involved in dealings of some sort with witchcraft. The one charge predominated in the south, the other in the north of Europe, as we should have anticipated. John Kepler, himself suspected of heresy, spent five years of his life, from 1615 to 1621, defending his mother from a capital charge of witchcraft. William

Witchcraft.

Harvey assisted at the examination of reputed witches. Giordano Bruno and Michael Servetus were burnt respectively by Catholic Rome and Calvinist Geneva for their religious opinions and their intellectual pre-eminence. The same penalty of death in the flames was reserved for both sets of victims. In each case, the civil authorities were only executing the decrees of the representative citizens of the day. The voice of the people was on the side of the powers whose actions condoned the popular alarms ; and Kepler's mother, after her acquittal, was preserved with difficulty from an indignant crowd.

It may be thought fanciful to connect witchcraft and witch-hunting with the advance of natural science or to associate them in any way with the prevalence of a mystical attitude of mind. Yet the evidence of racial susceptibility would accord with such a connection. Broadly speaking, the Northern races supplied the men of science, the witches and the mystics ; the East Anglian area, for instance, is remarkable for its contributions to all three categories.

The belief in witchcraft decayed with as little apparent reason as it rose. The civilized world, class by class, nation by nation, gradually discovered that it had ceased to believe in the existence of witches even before it had given up the practice of burning them. It was not that the world grew more tolerant or more humane, but that it had become more sceptical, and was preparing itself for the rationalistic philosophy and the intellectual tyranny of the eighteenth century.

To what was this change of attitude due ? Probably, to some extent, to the advance of science, which slowly defined the limits of man's mastery over nature and disclosed the methods by which this mastery is attained. Then, too, the extreme prepossession with theological matters, the natural result of the first popular contact with the material for the construction of religious belief, subsided as the years went on, and after a while the Devil and all his works ceased to perplex the mind of men. We now know more about these strange waves of unreasoning belief and religious exaltation which pass over nations, attracting especially within their sphere of influence the more ignorant, more neurotic, more highly strung of the population, and we may believe that the prevalence of a belief in witchcraft was, in certain aspects, an example, in an extended form, of such an epidemic.

But it is perhaps hardly wise to waive on one side, as the fancies of excited and ill-regulated brains, the whole of the phenomena which were classified under the general term of witchcraft during the heyday of the intellectual life of the Renaissance. Joseph Glanvill, a Restoration rector of Bath Abbey, who preserved for us the tale of the Scholar Gipsy made famous in Matthew Arnold's poem, found the evidence for some sorts of unexplained manifestations sufficiently strong to justify enquiry and examination. With Henry More, one of the band of Christian Platonist scholars at Cambridge who at that time exercised considerable influence in stemming the tide of naturalistic philosophy expounded by Hobbes and his school, Glanvill established what was virtually a small society for psychical research. He endeavoured in a manner worthy of the recently founded Royal Society, to which he belonged, either to confirm or disprove the phenomena under dispute. He investigated with some care a case of rappings at a Wiltshire house, in circumstances which recall the similar disturbances connected with the Wesley family. Again, associated in friendship with More, we find the " stroker," Valentine Greatrakes or Greatorex, who served in the Parliamentary forces under Roger Boyle, afterwards Lord Broghill and Earl of Orrery, and in later life found himself possessed of the power of curing many diseases, especially scrofula, by the touch of his fingers, accompanied with prayer. The efficacy of Greatrakes' ministry of healing, to which he devoted the latter part of his life, is testified to by Robert Boyle, the chemist, by Andrew Maxwell, by Benjamin Whichcote, and by many of the best-reputed men of the day. Two hundred years later, when another school of materialistic philosophy was in the ascendant in England, a second Society for Psychical Research was again founded by Cambridge men to investigate kindred phenomena, and a fresh outburst of healing by prayer and contact spread through certain sections of the community.

At the close of the seventeenth century Cambridge was the home of a group of idealist scholars and
The Christian divines of mystical tendencies, who were
Platonists. endeavouring to combine the new knowledge with the mysteries of the Christian faith. They felt that any merely mechanical explanation of the Universe, such as that attempted by Descartes, was unconvincing and inadequate in its essence, and expressed their conviction that the primordials of the world are not mechanical but vital. They developed

elaborate theories of the immortality of the soul, accepted some form of reincarnation, and believed in the possibility of communication between spirits existing in the æthereal, aerial and terrestrial states.

Human progress does not hold on its way in a steady line. The general trend of the curve may be upwards, as it certainly has been, in the matter of scientific thought, for the past four centuries. But superposed on the general curve are oscillations, often violent, which disturb its course.

Summary.

At the Renaissance, the movement towards rationalism in science became appreciable ; the triumphs of that spirit mark the years we have surveyed. The ideas of the French Encyclopædists, and the other tendencies of thought for the half-century preceding the Revolution, show the swing of the pendulum, which, as always, overshot its point of equilibrium, and, a century later, reached an extreme from which retreat was necessary by another path.

At the close of the Middle Ages, Thomas Aquinas gives the great example of a complete scheme of universal knowledge, including science, framed in accordance with orthodox Roman theology. From his day onwards we see a change. Duns Scotus and Occam at once began the separation between faith and reason necessary for the development of natural knowledge at that time. Yet the liberation of science from theological preconceptions was long delayed. The works of Descartes and Leibniz contain theological reasoning inextricably interwoven with physical facts. Even Euler (1744) based the principle of least action on the ground that the construction of the Universe is the most perfect creation possible, being the handiwork of an all-wise Maker, so that nothing can be found in the world in which some maximal or minimal property is not displayed.

It is true that in some of the greatest minds this confused thought is not found. Newton had a deep religious sense, and, indeed, wrote on theology. But his scientific work is quite free from theological arguments. Nevertheless, with less clear-sighted intellects, the theological prepossession gave an inward obstacle to scientific progress long after the danger of external persecution had passed away.

But a new tendency is seen in Locke and his followers. Aquinas built up a joint structure of knowledge, starting from the side of patristic theology. Locke, with characteristic British practical sense, and a wide experience of life and thought acquired at a critical period of history, essayed to found a rational Christianity on the sure ground of experience. Both attempted a synthesis. But while Aquinas' scheme had the rigidity and absoluteness of its chief constituent, Locke's contained the possibility of continual adaptation to the varying needs of intellectual development.

In the second half of the eighteenth century, the change of outlook became much more general. The ablest men in all branches of life became for the most part sceptical in matters of religion. Voltaire's attacks on the clergy and their teaching were but the most witty example of a wide-spread tendency of thought. Locke and the English Deists had their Continental counterparts, who undermined orthodoxy, just as the success of the Whig Monarchy in England tended to loosen the authority of legitimism in other countries.

Towards this general wave of sceptical thought the mechanical philosophy brought an important contribution. The astonishing success of the Newtonian theory in explaining the mechanism of the heavens led to an overestimate of the power of mechanical conceptions to give an ultimate account of the whole universe. As Mach says: " The French Encyclopædists of the eighteenth century imagined they were not far from a final explanation of the world by physical and mechanical principles ; Laplace even conceived a mind competent to foretell the progress of nature for all eternity, if but the masses, their positions, and initial velocities were given." Few would venture to hold such a sweeping conclusion nowadays, but, when first formulated, it was a natural exaggeration of the power of new knowledge which had impressed the minds of men with its range and scope, before they had realized its necessary limits. In fact, we have a repetition in changed circumstances of the story of the Greek atomists, who extended their successful speculative views of physics to the world of life and thought, all unconscious of the logical chasms which lay between—chasms only to be revealed and partially explored, but not bridged, by the tremendous accumulations of knowledge of two thousand years.

Another line of thought which reacted against the belief in the value of steady, unchanging dogma and authority, in religion and philosophy no less than in politics, was the growing idea of the progress of mankind in moral worth and intellectual power, and the not unnatural association of this progress with the achievements of the past few hundred years. Looking back to the mediæval times, it was impossible not to realize the immense step forward taken by the visible world of societies and nations in a comparatively short space of time. What was more natural to eager minds and generous impulses than first to imagine and then to believe that the process of development might be extended throughout the whole social body, that it needed but opportunity to enable all men to enter upon, share in and profit by the new inheritance?

Nowadays we know more about the limitations of the human mind, and we are perhaps more apt to dwell on them and on its imperfections than to dream of any yet unrevealed possibilities. To us, the marvellous epoch of the Renaissance is best interpreted as the first coming into free action of the intellectual forces of the Northern race; we share in the early triumphant breaking forth of the vanguard; we stand astonished at the extent of the conquest; we examine critically the details of the sober settlement. But as we read history, as we study more closely the great men of this great period, we realize that in truth but certain small sections of the social structure were affected, that only a very small and carefully prepared proportion of the whole population took active part in providing the festival. Of all the countries of Europe, five or six alone—Italy, France, Germany, England, the Low Countries and for a short time Spain—contributed appreciably to the intellectual sum total of the Renaissance. Again, in each country, the chief centres of activity are often obvious to us; and in

England, at any rate, we can almost circumscribe the portions of the nation, geographical and social, that ministered chiefly to the success of the whole movement. Thus we lose confidence in the idea of any general regeneration of the race by artificial means working from the outside, and recognize that there is some more fundamental, more elusive principle at the bottom of success or failure in scientific discovery as in all the other various walks of life.

It is difficult to realize that in the eighteenth century the idea of progress, still unproven perhaps, but familiar to ourselves, was strange and new. Scattered traces of the doctrine of the sustained and progressive advance of mankind are found in history " from Lucretius and Seneca to Pascal and Leibniz," but its first clear exponent seems to have been Turgot. When combined with the ideas of Rousseau about the natural equality of men, it did much to prepare the atmosphere of the French Revolution. It led to the conclusion, equally far from truth, that given favourable circumstances the people in the mass would prove necessarily as sincere, incorrupt, true and infallible as their greatest men—a conclusion indeed as old as Alcuin, though now first provided with a basis, demonstrably false to us, but convincing to certain minds in that age.

" The reign of reason," and the pathetic follies to which it led, are the real and tragic end of the period which began with the bright hopes of the Renaissance. After the French Revolution, we enter a new scientific age, which starts with the advantage of one more great disillusion, and a fresh determination on the part of the leaders of thought, as far as may be,

to see things as they are. The *a priori* methods of
the French Revolution were gradually left behind, to
consort with the yet older relics of mediæval scholasticism in the chilly confines of outworn, buried age.

In the period under review, not only did science
emancipate itself from theological prepossessions,
but it gained freedom from metaphysical trammels
also. This was a step in advance which the Greeks
failed to make, and it marked a change of outlook
necessary for success at that time in scientific thought.
Nothing is more striking than the difference between
the confusion of metaphysics and science which the
Aristotelian tradition imposed on the opening years of
the Renaissance, and the entire freedom from such
mixture with which the eighteenth century drew
towards its close. A general consensus of opinion in
fundamental scientific conceptions had removed their
subject-matter from the airy realm of philosophy to
the clearly defined territory of science.

At the beginning of the period, science had to conform to theology and to philosophy, in both of which
it was believed that something like finality had been
reached, in the first by the Roman Church, in the latter
by Aristotle. At the close of the epoch, science had
come into sure possession of its own new heritage,
and both theology and philosophy had realized that,
while still and for ever supreme each in its own em-
pyrean space, they must defer to the superior authority
of experience when they touched the firm ground of
natural science.

Pari passu with this separation of subjects, we get
a parallel separation of men of science on the one
hand from theologians and philosophers on the other.

The first requisite of the modern world as it emerged from chaos was authority, and it proved to a great extent that those who have the power of marshalling men have also the power of marshalling thoughts. The feudal system grouped men on an administrative basis, depositing authority with those strains of blood which had shown themselves competent to exercise it. The ecclesiastical organization was but one aspect of the art of government, but one method of organizing and directing the activities of the whole community. If the more ambitious and active members of the administrative classes usually found their scope in military and civil occupations, and established great families to perpetuate their hereditary talents, the more thoughtful, more acute, or more contemplative frequently entered the service of the Church, where at first learning of whatever kind, legal, medical and scientific, was concentrated, living in conditions of celibacy that probably circumscribed its development. Broadly speaking, the higher territorial families of England have supplied her with administrators, and the more numerous lesser gentry and yeomen furnished the doctors, lawyers and the men of science. Geographically speaking, the West and South of England sent the critics, the philosophers and the future administrators of England to Oxford, while the East Anglian area supplied to Cambridge the imaginative qualities, shown alike by scientific insight and poetical genius, in a long line of men of science and poets.

By the end of the eighteenth century, the social area from which most frequently the workers in science were drawn had begun to increase. It is an open question how far this change is due to the numerical expansion of the nation and its reorganization on lines of industrial efficiency, which would provide greater opportunities for intermarriage between the different sections of the community, and thus tend to enlarge, though possibly to dilute the quality of, production. Or it may be that science itself has changed, at least in one direction, and that the marvellous and successful application of its principles to industrial operations has required the co-operation of a new type of man, with a mind and body having an instinctive grasp of mechanical processes and technical skill wherewithal to apply the theoretical discoveries of the workers in pure science.

In the succeeding age, to which we must now turn, this process proceeds apace ; and science, hitherto with little message for the practical arts, unless therein we include medicine, opened wide the doors of technical application, and revolutionized the external circumstances of the life of mankind, just as, in the period we have just brought to a close, it had transformed the mental outlook.

CHAPTER V

The Mechanical View of Nature — The Popularization of Science — Astronomy—The Nebular Hypothesis—" Phlogiston "—Lavoisier and the Conservation of Matter — Imponderable Fluids — The Atomic Theory—The Wave Theory of Light—Spectrum Analysis—Heat and the Theory of Energy—The Rise of Electrical Science — Electric Waves—The Theory of Ions—Summary.

IF, in the story of knowledge, the French Revolution marked the end of the period which began with the

The Mechanical View of Nature.

Renaissance, yet many of the ideas which influenced the thought of the succeeding years were moving within its turmoil. Indeed the lines of advance which were to be characteristic of the new age had begun to appear here and there before that political cataclysm closed the eighteenth century. Thus, in the various branches of knowledge whose fortunes we pursue, our division cannot be made at a definite point of time. In every case we must turn back a few years to trace the origins of the threads we shall follow.

The broad tendency of the period now under review is to be sought in the gradual extension of the mechanical view of nature, which took its rise in the triumphs of the Newtonian astronomy as interpreted chiefly by French mathematicians, to other branches of know-

ledge, till, in the Darwinian theory of natural selection, its less clear-sighted advocates saw a complete
account of life and its phenomena.   In a future
chapter we shall trace the widening of this sharply-
cut but somewhat dry and meagre philosophy into a
broader and fuller idea of the complexity and wonder
of the Universe.   But, throughout the greater part of
the nineteenth century, the true method of advance
lay in building up this naturalistic edifice in new and
extending directions, and in proving its capacity to
interpret phenomena, of which some were thought
inexplicable and others were then unknown.

For the work of the Renaissance and the succeeding
ages had been to bring within the circle of the com-
prehension of man phenomena with which he had
long been acquainted, and most of which were easily
apparent to his unaided senses.   But now we pass
to a time when scientific work consisted largely
in revealing the hitherto unknown and hitherto
unknowable as well as in explaining them when revealed.   An extension of the senses had been produced
by the perfectioning of such instruments as micro-
scopes and telescopes, and their adaptation to photography.   Hence followed the discovery of living cells
as the units of the organism, and of the structure of
far-distant stars and suns.   Then the disclosure of
all the vast and complex phenomena of electricity,
to name but the most striking instance out of many,
opened up fresh worlds for science to conquer, worlds
which proved amenable to the same logical methods
of investigation that were first tested in dynamics
and astronomy.

When the nineteenth century opened, the great

industrial development which is still in progress had already begun. One of its chief instruments, the steam engine, reached a serviceable form when, in 1769, James Watt patented the principle of the condenser. The steam engine was a practical invention, to which scientific principles were applied at a later stage to carry out developments and improvements. But the electric telegraph, the other great agent in revolutionizing the social conditions of the world, was a direct consequence of research in pure science.

To some people, the practical applications of science stand for its main achievement. But their effect on the human mind and its thought, though great, is indirect. That effect is slow and cumulative. The gradual and apparently inevitable extension of man's power over the material resources of nature gives applied science, by which the advance is chiefly secured, an importance in the eyes of the outside world with which no amount of abstract thought endows pure knowledge. Indeed, in the eyes of the world, applied science attains the position which Francis Bacon foresaw and desired. As one triumph after another is won, the effect to all appearance is that of an invincible if slow advance. It seems that no limits can safely be assigned to the extension of man's mastery over nature; it comes to be assumed that the mechanical principles by the application of which that extension is made are competent to account for and explain the whole of the Universe.

Save in this indirect way, with the technical applications of science we shall have little to do. It is indeed a very grave question whether the consequent advance in material wealth and resources has produced any corresponding elevation of tone or deepening of insight in the human mind. Sometimes it looks as though one effect had been an actual lowering of ideal, an increase in bodily servitude and spiritual despair. We have come to regard physical comfort and personal security as the main object of human endeavour, and have forgotten the old truth that he who would save his life must be ready to lose it.

Even on the material side, a rapid advance in the means of exploiting the resources of Nature has brought many perils. In the early years of the nineteenth century, an enormous increase of available energy, produced by the new means of drawing on the limited and stored supplies in the coalfields, was brought within reach by the mechanical inventions of a small number of able men. This energy gave the means of supporting a larger population, and at once the population rose in response. But organization, social, civil and religious, failed to keep pace with the industrial changes, and to adjust itself to the new proportions in which the grades of this rapidly growing population became divided. The tremendous increase of wealth fell largely into irresponsible hands, and the directing and controlling power of the State, at any rate for a time, became almost indistinguishable in the medley of conflicting interests. Thus it is that, as the period we are now studying draws on, as a result of the industrial development and the unstable social conditions it created, we shall find men driven to study their corporate history and their social maladies by methods learned in biological and statistical science, and discovering facts and formulating problems undreamed of by their predecessors. In this latest field added to its estate, science has set itself to examine experimentally the properties and needs of the social organism, and to deal with the needs of the human mind as well in its collective as in its individual aspect.

But, during the greater part of the period under review, the complication of the social problem was not understood, and social questions were answered in the same confident spirit which was manifested in other branches of science. Thus the cumulative effect of the general tendency, in pure science, in its technical applications, and in social philosophy, became very great. It seemed that the human mind was fast coming to a complete explanation of all things on a mechanical and materialistic basis.

The diffusion of this limited and one-sided aspect of Nature, unaccompanied by historical or critical insight, among minds unsuited by character and education to apprehend the underlying deeper problems, produced the narrow and dogmatic type of scientific thought prevalent during a great part of the nineteenth century and generally associated with a materialistic outlook on life. A knowledge of the history of science, an appreciation of the inadequacy and temporary nature of many of its hypotheses which have done good work in their time, together with the realization of the deeper metaphysical questions which lie all unanswered beneath science at every point, are tending to release the human mind from the iron domination of nineteenth-century scientific scholasticism, which was threatening to outlive its period of usefulness as a corrective to the older dogmatism of the modes of thought it superseded.

Throughout the century, it is true, we find that most of the great leaders in science, both in their own personalities and in the general trend of their teaching, keep their touch with the deeper realities of the unsounded depths of the human soul. They, at all events, still grasp, unconsciously it may be, the connection between the experimental method in natural science and an attitude of open-minded reception of spiritual experience ; between the intuitive religious instinct of the great mystics and the scientific insight of a Leonardo or a Faraday, who apprehend intuitively the essence of a problem and frame conceptions fit to guide not only their own experiments and those of the lesser men who follow, but also to throw light on the meaning of life itself. Such men as Faraday, Pasteur, Stokes, Kelvin (who is said to have begun his lectures on physics with the Collect for the day) were very far from the materialistic dogmatism of some of their followers. This humility, this sane and calm balance of mind, is oftener found in great physicists who deal with subjects of the more fundamental kind, subjects in which such conceptions as matter and force, in which other sciences are apt to see complete and conclusive explanations of natural phenomena, are themselves analysed and, as ultimate verities, found wanting. It is less common among those who have to apply the more fundamental concepts to the secondary sciences, nor is it often developed among those whose business it is to apply the results of science to the arts of practical life. An overestimate of the function and power of science becomes as common in one direction as an ignorance of the breadth and scope of the ground it may rightly claim is in another.

The social philosophy of the nineteenth century bore marks of the hopes of the Encyclopædists for the explanation of the Universe, and of the French Revolution for the regeneration of mankind. Once again immense strides were taken ; once again the forward movement gradually became overshadowed with the certainty of another great disillusionment for those who had mistaken the means for the end.

Like the Israelites in the wilderness, the modern world cried out to science to make it a golden calf. The golden calf was duly made and greatly admired, for it had its uses ; but it has been found wanting as an object of worship. Meanwhile, new tables of the law are being delivered to the prophets of pure knowledge, and a fresh revelation is forthcoming of the terms of the covenant by which alone it is possible for human nature to fulfil the immediate purpose of its existence in the age now beginning.

Newton and his contemporaries wrote in Latin. Their works were thus the common property of that small section of each nation that understood the ancient tongue—sections that were probably akin to each other in race as well as in modes of thought. The very limitation and detachment inseparable from publication in a classical language produced advantages to set against its drawbacks. Science was cosmopolitan ; it was apart from the national life ; but was thereby saved from the dangers of popular interference, and the degradation of popular approval of the material benefits it may confer.

The Popularization of Science.

When we take up the tale once more, in the latter

half of the eighteenth century, there has been a
general change to the vernacular in scientific literature.
Science thus became part of the general intellectual
development of each people.   The change did much
to make possible the technical application of scientific
results and the industrial upheaval of the nineteenth
century.   But it came at a period when a materialistic
view of science was in the ascendant, and the nations
of Christian Europe, as a whole, received as their first
impression of the new knowledge the idea that the
object of scientific enquiry is the creation of wealth
and the provision of practical benefits—an attitude of
mind that is still widely prevalent in our midst—and
that a mechanical explanation of existence is possible
or even has been accomplished.

The popularization of science was a slow process.
At first, in the seventeenth century, the foundation
of the Royal Society of London and of similar societies
in other capitals gave a focus and meeting-place for the
men of science within reach.   During the succeeding
century the great extension of interest in scientific
matters led to the foundation of various provincial
societies which had considerable influence in the spread
of the new knowledge.   The Royal Society of Edinburgh, which still survives, was founded in 1783 on
the framework of an older Philosophical Society.   The
Lunar Society flourished in the Midlands of England
at the close of the eighteenth century, and brought
together such men as Erasmus Darwin, S. J. Galton,
and Priestley, who rode across country at each full
moon to the appointed rendezvous, to discuss the
scientific problems of the hour.   These local societies
for the most part decayed when the facilities of railway

locomotion drew all such interests to London or the two Universities, where, for fifty years, learning in England became chiefly concentrated. It is only of recent years that the foundation of provincial Universities has re-established local centres of interest and intercommunication.

The first line of scientific achievement we have to follow into the new era is an extension of old methods rather than an invention of new ones.
Astronomy. Newton's *Principia* carried the gravitational theory as far as the existing mathematical tools allowed ; but during the years 1773 to 1787 that theory was developed much further by the use of modern mathematics by J. L. Lagrange and Pierre Simon, afterwards created Marquis de Laplace. In Laplace's second great period of activity, 1799–1825, he systematized all such knowledge in his monumental treatise, *Mécanique Céleste*, and popularized the Newtonian philosophy in a smaller book, *Système du Monde*.

Almost all the complicated planetary motions were explained broadly by this school of mathematical astronomers, though some details of lunar and planetary theory were left to be worked out in future years. It may be said at once that the final test was given to Newton's theory by its use to predict the existence of an unknown planet, thus reversing the methods of Newton, Lagrange and Laplace. The perturbations from its orbit of the planet Uranus were not to be accounted for fully from the action of the other known bodies, and, to explain these irregularities, the influence of a new planet was assumed, and its necessary position

11

calculated independently by John Couch Adams of Cambridge and the French mathematician Leverrier. Turning his telescope to the position indicated by Leverrier, the astronomer Galle of Berlin detected a planet to which the name of Neptune was given (1846). These researches give the final verification of Newton's fundamental hypothesis that each particle of matter in the solar system may be supposed to attract every other particle with a force proportional to the product of the masses, and inversely proportional to the square of the distance between them. The probability in favour of the hypothesis as the embodiment of a correct relation is possibly as great as that in the case of any other physical theory hitherto propounded.

Newton's theory, as developed by himself, Lagrange and Laplace, proved sufficient to co-ordinate all the observations made concerning the solar system as now in existence ; it was therefore natural to try to extend its sway over the past, and to adapt it to an attempt to explain the origin of that system. Hence Laplace framed a " nebular hypothesis," which pictured the primordial chaos as filled with scattered matter like a diffused cloud, spread throughout the space now occupied by the solar system. Laplace showed that known dynamical principles were consistent with the drawing together and gradual solidification into distinct fiery masses of such space-scattered particles. He thus made conceivable to the human imagination the formation of the sun and planets from a formless chaos. The existence of luminous nebulæ in the

The Nebular Hypothesis.

heavens, fulfilling the required condition of vapour-like diffusion, is a well-ascertained fact, and the nebular hypothesis gains rather than loses support from recent astronomical research. The French Encyclopædists, earlier in the eighteenth century, had begun to think that they could see their way to explain all nature on mechanical principles, and, as we have said, Laplace imagined " a mind competent to foretell the progress of nature to all eternity, if but the masses, their positions and initial velocities were given." This may be regarded as the complete view of thorough-going mechanical philosophy, which gained increasing hold over the scientific world in the period now under consideration. Its confident completeness is well illustrated by Laplace's reply to Napoleon's query about the place of God in the scheme of nature, " that he had no need of that hypothesis."

But while mechanical science developed in paths already mapped out, chemistry underwent an entire transformation. Thus, at the beginning "Phlogiston." of the period, the Aristotelian conception of a body essentially light kept its hold in Stahl's hypothetical substance " phlogiston." Misled by the phenomena of flame, which naturally suggest an escape of something from the burning body, Stahl framed a theory in which that something was accepted and named " phlogiston." Since, when all the products are collected, the balance shows a gain, phlogiston must possess a negative weight. In terms of this hypothesis chemical science had learnt to express its facts, and, owing to its influence, isolated investigations, which pointed to more modern views, had failed to impress the minds of chemists, and had to be rediscovered when time had gradually undermined the theory and prepared the way for the natural reinterpretation of the phenomena. In 1669, a century before Priestley's final discovery of oxygen, its existence in air and its significance in the phenomena of respiration and combustion had been demonstrated by John Mayow (1640–1679), a physician who practised in Bath and London. Again oxygen was prepared from heated saltpetre by Borch in 1678, and once more in 1729 by Hales, who actually collected it over water. The isolation of hydrogen may even be traced back to Paracelsus, who described the action of iron filings on vinegar. Yet all these observations were forgotten and their meaning lost ; air was still believed to be the only ponderable gaseous element.

Thus once more we are impressed by the rarity of finding that the accepted discoverer of a scientific phenomenon, or the received originator of a scientific theory, stands alone or even at the beginning of the episode. A study of previous records nearly always discloses others who trembled on the verge of the discovery, or beheld premonitory visions of the coming revelation. Leonardo da Vinci buried in his note-books many of the ideas which created the science of the succeeding centuries—ideas which only attained an accepted position in the structure of knowledge when Galileo or Huygens had fitted them into place. Behind Leonardo, again, stand other shadowy figures of whom we catch here and there a faint glimpse. It is only when the frame is ready, when a place is waiting for the new conception, that it can be certain of an immediate and enduring recognition. A discovery or a theory that is too much before its age stands but a poor chance in the ordeal of selection. Of many men's work it may be said, as Kant said of his own writings, that they had come a century too soon.

The beginning of the change appears in the work of Dr Joseph Black of Edinburgh, who about 1755 discovered that a new ponderable gas, distinct from atmospheric air, was combined in the alkalies. This gas he named " fixed air " ; it is what we now call carbon dioxide or carbonic acid. Nevertheless, for yet a time phlogiston lived.

In Joseph Priestley (1733–1804), theologian and experimenter, we find the man who stood at the threshold of the new era. He prepared oxygen by heating mercuric oxide, and discovered its marked power of supporting combustion. But he described it as dephlogisticated air, and failed to perceive that his discovery turned a new page. As Cuvier said, he was the father of modern chemistry, but never acknowledged his daughter. Priestley had the distinction of having his critical history of the corruptions of Christianity burned by the common hangman at Dort, and his papers and experimental laboratory destroyed by a Birmingham mob.

The theory of phlogiston maintained its hold also on Henry Cavendish (1731–1810), a grandson of the second Duke of Devonshire, who inherited alike his scientific ability and the means wherewith to develop it from a family where both attributes are not uncommon. Cavendish made many discoveries, especially in electricity, which have been credited to other men. His retiring and secretive disposition led him often to neglect the duty of publishing to the world the results of experiments made solely to satisfy his own curiosity. This was not the case with his chief chemical discovery, the compound nature of water. But when, in 1781, Cavendish thus dethroned water from its old and proud position as one of the elements, he still described its constituents as phlogiston and dephlogisticated air.

But, with this gradual accumulation of knowledge, the accepted theory was becoming more and more inadequate. New ideas were better able Lavoisier and the Conserva- to describe the phenomena, and, in due tion of Matter. time, the new ideas appeared. In 1783, Lavoisier repeated Cavendish's experiment, and grasped the fact that there was no need to invent a body with properties fundamentally unlike those of other material substances. Lavoisier regarded the constituents of water as ordinary gases possessing mass and weight, and named them hydrogen (the water-forming element) and oxygen (the acid-forming element). The conception of phlogiston became unnecessary, and with it vanished the last of the essentially light bodies with negative weights. Thus the principles which, for a century and a half, had revivified the science of mechanics were carried over into chemistry.

In Lavoisier's hands, too, another principle which had slowly been emerging from obscurity became clear : the principle of the persistence of matter, or conservation of mass. Throughout any series of chemical changes, the matter involved may alter its state of chemical combination, and change its form from solid to liquid or seemingly disappear as a gas.

Yet, when collected and tested by the balance, the products of the reaction weigh the same as the reagents. This principle is accepted by Lavoisier in all his investigations ; he states it in definite form : " The quantity of matter is the same at the end as at the beginning of every operation."

Nevertheless, in spite of the disappearance of phlogiston, and the establishment of the principle of Imponderable the conservation of mass, the conception Fluids. of a weightless fluid had still many years of useful life before it. Although the most acute of the natural philosophers, such men as Newton, Boyle and Cavendish, inclined to the opinion that heat was due to a vibratory agitation of the particles of bodies, in the absence of definite conceptions corresponding to our modern notions of energy, that opinion could not be developed into a useful working hypothesis. The advance which was waiting to be made was the idea of heat as a measurable quantity, unchanged in amount as it passed by contact from one body to another. To experiment with this conception as a guide, men needed a definite and suitable representation of the nature of heat. This conception was at hand in the theory that heat might be a subtle, invisible, weightless fluid, passing between the particles of bodies with perfect freedom. It was in the light of this theory that Black discovered and investigated the phenomena of the latent heat required to produce a change of state from solid to liquid, or liquid to gas, and the specific heat needed to raise the temperature of a substance. He thus established the science of calorimetry, or the measurement of a quantity of heat.

A similar theory, or rather two similar rival theories, enabled Franklin, and his fellow-workers, to throw light on the phenomena of electricity, which were now drawing men's attention.   The attractions and repulsions between electrified bodies may be expressed on the supposition that there is a substance, electricity, which, like heat, may be treated as a quantity subject to the laws of addition and subtraction.   With electricity, however, the existence of two distinct and opposite varieties was revealed in an early stage of the history of its discovery.   An electric charge developed by friction on glass will neutralize a charge produced in the same way on ebonite.   These results were explained by the supposition either of two fluids with opposite properties, or of one fluid, of which the excess or defect from the normal quantity gave rise to the electrified state.   The terms of speech appropriate to the one-fluid theory, with its positive and negative electricities, are still with us, though, as we shall see later, the fluid has given place to the newer conception of electric particles or corpuscles, or to the vaguer one of disembodied charges or electrons, the excess of which constitutes the conventionally negative, and the defect the conventionally positive, electrification.

The overthrow of the theory of phlogiston had the effect of bringing clearly to light the three states of The Atomic matter, solid, liquid and gaseous, as Theory. we know them now—different physical phases of the same substance, which primarily may be best known in any case in one of the phases, but in suitable conditions can exist in either of the three. This advance in knowledge was followed by the study of the laws of chemical combination, since gases, in which the laws can be traced most simply, had been now brought into a correct relation with other bodies.

As the result of careful analysis, it was found that a chemical compound was always made up of precisely the same amount of the constituent parts, and this fact of the fixity of composition played an essential part in the scheme of the new chemistry. Water, however obtained, always consists of hydrogen and oxygen combined in the ratio of one to eight. Thus the conception of combining weight was reached, the combining weight of oxygen being eight, if that of hydrogen be taken as unity. Once again, if two elements combine in more than one way, to form more than one compound, the proportion of the constituents in one compound is simply related to the proportion in the other : fourteen parts of nitrogen combine with eight of oxygen in one compound, and with sixteen parts in another, exactly double the first proportion.

John Dalton (1766–1844), a colour-blind Quaker chemist, who was born in Westmoreland and worked and died at Manchester, saw that these phenomena pointed to the conclusion that matter was not infinitely divisible, and that combination took place between discrete particles with definite weights characteristic of each substance. Led by these considerations, he revived the old atomic theory in modern guise, the essence of the new form of it being that it was possible to arrive at a knowledge of the relative weights of the atoms of two elements from a determination of the proportions in which those elements combine in their simplest compound. Thus the composition of water indicates that the atom of oxygen is eight times as heavy as the atom of hydrogen, and that the constitution of water should be represented as HO, where H stands for one atom of hydrogen and O for one of oxygen. By such methods Dalton assigned atomic weights to some twenty elements (1804).

The insufficiency of Dalton's conceptions as they stood became apparent when the phenomena of gaseous combination were studied more extensively. Gay-Lussac (1778–1850) showed that gases always combine in volumes that bear simple ratios to each other, and Americo Avogadro, Conte di Quaregna (1776–1856), pointed out that on Dalton's theory it followed that equal volumes of all gases must contain numbers of atoms that bear simple ratios to each other. But it was soon found, both from further study of gaseous combination and from physical considerations, that a distinction was necessary between the chemical atom, the smallest part of matter which can enter into combination, and the physical molecule, the smallest particle which can exist in the free state. The simplest method of expression of Avogadro's hypothesis is to suppose that equal volumes of gases contain the same number of molecules. The physical theory developed in mathematical form by Waterston and Joule, that the pressure exerted by a gas is due to the impacts of the molecules, which are in a state of perpetual movement and collision, also leads to this result.

But, to return to the case of water, two volumes (and therefore molecules) of hydrogen combine with one of oxygen to form two volumes (or molecules) of steam. It will be seen that the simplest theory which will explain these relations is one which supposes that the physical molecules of hydrogen and oxygen each contain two chemical atoms, and that the molecule of water vapour has the chemical composition represented by H_2O. Thus Dalton's combining weights need to be brought into line with the facts revealed by the later experiments before we are in a position to assign to the elements their true atomic weights.

Dalton's theory, when developed by the addition of Avogadro's hypothesis, arose naturally out of definite experimental knowledge. Other phenomena were waiting to be interpreted by its light, and still more were discovered and co-ordinated as time went on. Except for a temporary tendency in the last few years of the nineteenth century to replace the theory of chemical atoms by relations based on the theory of energy, it may be said that atomic and molecular conceptions have inspired most of the chemical and physical research which has distinguished the past century.

One result of Dalton's work was to give a working definition of an element, as a substance possessing a distinct and unchanging atomic weight from whatever source it was obtained. Hence, since it was now possible, it became of high importance to extend the knowledge of the substances which were in truth elements, incapable of further analysis or subdivision.

The number of known elements has grown from the twenty recognized by Dalton till now some eighty different kinds of matter have been recorded. The work of discovery has proceeded fitfully. When any new method of research has been applied to chemical problems, a new group of elements has frequently come to light. The separating power of the galvanic current enabled Sir Humphry Davy to isolate the alkaline metals potassium and sodium in 1807. At a later date, spectrum analysis showed the existence of such substances as rubidium, cæsium, thallium and gallium ; while, in recent years, the still more delicate methods of radio-activity have disclosed bodies like radium and its allied family of elements.

The atomic weights of the elements, first calculated by Dalton, are constants of fundamental importance in natural science. Relations between the values for different elements which possessed somewhat similar properties were pointed out by Prout and Newlands, and in 1869 the Russian chemist Mendeléeff showed that, if the elements be arranged in a table in order of ascending atomic weight, there is a remarkable likeness in the physical and chemical properties of various groups of elements as they fall at certain regular intervals in the table. An evident gap in the series, which could not be filled without introducing confusion into the scheme, even led Mendeléeff to prophesy the existence and to predict the properties of an element which was not discovered till some time afterwards.

This " periodic law " shows that the physical and chemical properties of the elements depend on the mass of their atoms. And, since it also thus makes clear the existence of some relation between the atoms themselves, it leads us back once more to the old idea that different substances are made up of the same fundamental stuff, with its ultimate particles differing in arrangement or in number in the different atoms. Here the required penetration into natural phenomena grows too intense for the methods proper to chemical research, and the actual discovery of an ultimate particle, common to all kinds of matter, is one of the latest triumphs of physics to be described below.

Meanwhile a vast extension of chemical knowledge was made by optical methods. As we have already seen, Newton preferred the emission to the undulatory theory of light because of the difficulty of explaining the phenomena of straight rays and shadows on any hypothesis of wave motion. But in 1801–3 Thomas Young revived the wave theory to account for his newly discovered phenomena of the interference of two rays of light to form coloured fringes, as well as the colour of thin plates studied by Newton. Young's ideas were developed in mathematical form by Fresnel, who also showed that rectilinear propagation would be a property of waves which were very small compared with the dimensions of the obstacles or the distances concerned, and thus met Newton's objection to the theory. Moreover, the phenomena of polarization, which indicate that a ray of light possesses different properties on different sides of the ray, were shown by Young to prove that the direction of the vibrations which constitute the wave is at right angles to the direction

The Wave Theory of Light.

of propagation. Finally, since the hypothesis of waves certainly connotes the idea of something to vibrate, it was thought necessary to revive the old conception of an all-pervading universal æther to supply what the late Lord Salisbury once called " a nominative case to the verb ' to undulate.' "

The wave theory gave an explanation of the phenomena which led to the triumphs of spectrum analysis. Joseph Fraunhofer (1787–1826) had mapped the black lines which cross the coloured band of light, or spectrum, which is obtained by passing a beam of parallel rays of sunlight through a glass prism. The nature of these lines seems first to have been made plain by Sir George Gabriel Stokes (1819–1903) in his lectures at Cambridge, though with characteristic modesty he gave his ideas no wider publicity. A child's swing is set in motion by giving it a series of small impulses which coincide with its natural period of oscillation, and, in similar manner, any mechanical system will absorb energy which falls on it in rhythmic unison with its own natural vibrations. Hence the molecules of the vapours in the outer envelope of the sun will absorb the energy of those particular rays coming from the hotter interior, of which the oscillatory period coincides with their own. The light which passes on, then, will be deprived of light of that particular colour (i.e. frequency of vibration), and a black line in the solar spectrum is the result. Bunsen and Kirchhoff saw that, if this explanation be the true one, the black lines due to absorption should correspond in position with the bright lines the molecules of the same element themselves give out when heated. Thus, when one of Fraunhofer's lines coincides with the bright line in the laboratory spectrum of a terrestrial substance, the presence of that element in the envelope of the sun is to be inferred. Moreover, by passing the intensely white light of the electric arc through sodium vapour, volatilized in the cooler flame of a spirit-lamp, it should be possible to reproduce artificially the dark absorption line of sodium. Such an experiment indeed had been carried out by Léon Foucault in 1849, but had passed unnoticed and been forgotten. It was performed once more by Bunsen and Kirchhoff in 1860, in the light of the theory given above, and its success at once opened up the new science of the chemistry of the sun and stars. By analysis of their light, it became possible to detect the existence in the depths of space of elements known on our globe. Moreover, since the colour of light and the relative position of spectral lines depends on the frequency of the vibrations which reach the eye, a luminous source approaching an observer will crowd more waves into his eye per second than if the source were at rest. Thus a slight shift, in one direction or the other, in the position of spectral lines which can be measured microscopically indicates relative approach or retreat between the source of the light and the earth. In this way the velocities and movements of stars can be investigated by a study of the light which they emit. Although many stellar spectra are too faint to affect the eye, in photography small effects can be made cumulative, so that, by long exposure, lines invisible to the unaided eye can be detected on a photographic plate.

As we have seen, the caloric theory, by which heat was pictured as an imponderable fluid, played a useful part in suggesting and making intelligible experiments on the measurement of quantities of heat.   But, as a physical explanation, it had always seemed insufficient to the more acute natural philosophers.   Investigations by Count Rumford on the boring of cannon, and by Sir Humphry Davy on the rubbing together of two lumps of ice, had shown that heat was developed by friction to an unlimited extent in a way inexplicable on the caloric theory.   But more direct experimental proof was needed.   James Prescott Joule, during the years between 1840 and 1850, measured accurately the amount of heat produced by the expenditure of a known quantity of work, and proved that, however the work was expended, the same amount of heat was produced.   When these experiments became known, an exact equivalence between heat and work was generally recognized, and heat accepted as a form of motion.   It was now regarded as the average energy of motion of the molecules of a body, and when, as in a gas, the molecules are supposed to be free from each other's influence, it is easy to obtain a vivid physical conception of the state of affairs.   Owing to momentary collisions, at any instant there will be molecules moving with any number of different velocities in all sorts of directions.   The impact of the molecules on the walls of the vessel constitutes the pressure of the gas, while the average energy of the molecules measures its temperature.   On these lines Joule gave an elementary mathematical kinetic theory

Heat and the Theory of Energy.

of a gas, which was extended and made more rigorous by James Clerk Maxwell. On this theory, most of the physical relations of a gas can be explained directly in terms of simple movements of the hypothetical molecules of which it is imagined to be composed.

Joule's definite experimental result that work and heat were equivalent gave point and power to a more general conception which was then arising, known as the " correlation of forces," and developed it into the important physical principle of the conservation of energy. Energy, as a definite physical conception, was new to science. The idea which underlay it had previously been expressed by an inaccurate and confusing extension of the meaning of the word force. Energy may be defined as the power of doing work, and, if the conversion be complete, may be measured by the work done.

Joule's experiments showed that, in one definite case, the total amount of energy in a system is constant, the quantity lost as work reappearing as heat. General evidence led to the extension of this result to other changes, where, for instance, mechanical energy is converted into electrical energy, and then perhaps into chemical energy. All known facts are consistent with the statement that, in all cases in present conditions, the total energy of an isolated system is constant in amount.

The principle of the conservation of energy, thus established, is comparable with the long-known principle of the conservation of mass. Newtonian dynamics are founded on the recognition that there is a quantity, for convenience called the mass of a body, which remains constant throughout all motion. In the hands of the chemists, the balance showed that this principle held good also when chemical changes occurred. A body burning in air was not annihilated. When the resultant gases were collected, the total weight of body and air remained unaffected.

And so it is in the case of energy. We find another quantity besides mass which remains unchanged throughout a series of transformations. Hence it is convenient to recognize the existence of that quantity and to give it a name. We call it energy, and measure its changes by the amount of work done or by the amount of heat developed.

As we can neither create nor destroy matter, so we can neither create nor destroy energy. We can but convert matter or energy from one form into another.

The importance of this result in the history of physical science was immense. It gave a new method of investigation, a new point of view from which physical problems could be regarded. Since the final energy of a closed system must be the same as that at the beginning, it became possible in some cases to predict the final state of the system without reference to intermediate steps, to pass at once to the solution of a problem without tracing the process by which the goal was attained. As a practical guide in scientific investigation, the principle of the conservation of energy is one of the great achievements of the human mind.

But, if the principles of the conservation of matter and energy hold good in all the circumstances we can investigate, and to all the accuracy we can attain by experiment, it was natural to stretch the principles into the form of general laws. Matter became eternal and indestructible ; the amount of energy in the Universe became constant and immutable, in all conditions and for all ages. The principles passed from safe guides for empirical advance in knowledge, into philosophic dogmas of doubtful validity.

When heat is transformed into work, or work into heat, the equivalence between them is completely expressed by Joule's results. But, although it is always possible to transform the whole of a given quantity of work into heat, it is not generally possible to perform completely the reverse change. In steam engines, and other heat engines, it is always found that only a fraction of the heat supplied is transformed into mechanical energy ; the remainder, which passes from hotter to colder parts of the system, does not become available for the performance of useful work. Every heat engine needs a difference of temperature, a hot body or source of heat and a cold body or condenser. The possible efficiency of the engine increases as this difference of temperature increases, and, in a theoretically perfect frictionless engine, the efficiency is independent of the form of the engine or the nature of the working substance, and gives a clue to an absolute scale of temperature which also is independent of the properties of any one substance.

On these lines the principle of the relations between heat and work, the science of thermodynamics, was established, chiefly by the labours of Sadi Carnot (1796–1832), William Thomson, afterwards Lord Kelvin (1824–1907), and R. J. E. Clausius (1822–1888).

The consequences of thermodynamic reasoning have not only enabled the engineer to place on a firm footing the theory of the heat engine, but have aided materially the progress of modern physics and chemistry in many other directions. They led to the theory of an absolute scale of temperature, and pointed the way to that series of researches at low temperatures which eventually liquefied all known gases and completed the proof of the continuity of all types of matter in its three states.

For the gain of useful work from a supply of heat, a temperature inequality is necessary. But, in nature, temperature inequalities are continually being diminished by conduction of heat and in other ways. Hence, in a closed system, the heat energy tends steadily to become less and less available for the performance of useful work. When the available energy becomes a minimum, no further work can be done, and thus we determine the necessary conditions of equilibrium of the system. In this way the mathematical theory of chemical and physical equilibrium has been built up by Kelvin, Helmholtz and Willard Gibbs.

Similar ideas, extended to the whole solar system, indicate that mechanical energy is continually wasting into heat by friction, and heat energy continually becoming less available by the reduction of inequalities of temperature. Hence we are led to contemplate a distant future in which all our stores of energy shall have been converted into heat uniformly distributed through matter in mechanical equilibrium, and all further change becomes for ever impossible. But it should be noted that this conclusion rests on several unproved assumptions. It supposes that the solar system, or the Universe if it be extended thereto, may be treated as an isolated system into which no energy is entering ;

it supposes that individual molecules, the velocity of which is subject to continual alteration owing to collisions, cannot be followed and separated into fast-moving and slow-moving groups, for, as Maxwell pointed out, such a power would enable diffused energy to be reconcentrated.   Nevertheless, in the conditions of Nature as then known, the conclusion stood ; the energy by which we live and move was seen to be becoming continually less available, the iron hand of mechanical necessity stretched out to grasp and crush the life of the Universe.

The steam engine was developed by practical men to meet industrial needs, and owed little, except in its later stages, to the help of pure science.   But, as regards the other great physical agency of modern civilisation—electricity—the case was very different.   Here the practical applications followed, and were a direct consequence of, the researches of men who sought for knowledge alone.   Modern electrical science began with the discovery of the phenomena of " galvanism " or current electricity by Galvani in 1786, and Volta of Pavia in 1800.   Volta's cell, from which he produced the first electric current, consisted essentially of two plates of unlike metal, placed in the solution of an acid or a salt.   This apparatus was found to show effects which were soon correlated with those of the older frictional electric machines.   While the frictional machine supplied isolated charges of small quantities of electricity at a very high tension, showing their existence on discharge by visible sparks, the new voltaic battery gave a steady large current at a much lower

The Rise of Electrical Science.

electromotive force. The chemical decompositions produced by these currents were the first of their properties to be investigated, and led to the industry of electro-plating, while, more recently, they have been used extensively in many processes of chemical manufacture and in metallurgical operations.

In theoretical science, one of the earliest and most striking of the results of the use of the voltaic cell was the separation in 1807 of the alkali metals potassium and sodium from their oxides in the researches of Sir Humphry Davy (1778–1829). Davy, a native of Penzance, was the first Professor of Chemistry at the newly founded Royal Institution. It is a sign of the growing general interest in science that an institution for diffusing scientific knowledge should have come into existence, and that Davy's lectures should have attracted a large audience composed both of " men of science and numbers of people of rank and fashion." Davy's attractive personality had much to do with this success, while his imaginative and poetic nature was an essential part of his power as a framer of scientific hypotheses fit to be tested by experiment.

Electrically, Davy's work marks an important step. The use of electrical forces to separate chemical compounds into simpler constituent parts indicates a connection between electrical and chemical phenomena which is of fundamental importance in the theory of both subjects. Davy's work was taken up again and carried further by his pupil and successor, Michael Faraday (1791–1867), whose instinctive genius we shall meet again in future pages. Faraday showed that the amount of chemical action was strictly proportional to the quantity of electricity which passed, and that the mass of substance separated was proportional to its chemically equivalent weight.

The great industrial applications of electricity, made possible by the invention of the telegraph on one hand and the dynamo on the other, are founded on the scientific discoveries of two physicists. In 1820 Hans Christian Oersted of Copenhagen discovered that, if a wire be placed above or below a magnetic compass needle and parallel with it, the needle is deflected from its north-and-south position when an electric current is passed along the wire. By the construction of galvanometers, this magnetic force was at once applied to the measurement of currents, and underlies all our modern quantitative science of electro-magnetism. In 1831 Michael Faraday demonstrated at the Royal Institution that, when the current in one circuit of wire is started or stopped, a momentary secondary current is induced in another coil of wire placed near the first one. A similar current is produced by the motion of magnets, or by any other change in the magnetic force in the coil. These currents, barely perceptible to Faraday's primitive instruments, were set up by means the same in principle as those which now drive our electric trains and supply thousands of horse-power to many industrial manufactures.

Faraday's experimental researches were guided by his instinctive grasp of the importance of the dielectric
Electric Waves. or insulating medium in electric phenomena. When a current deflects a magnetic needle across space, or induces another current in an

entirely unconnected circuit, we have either to suppose an unexplained "action at a distance," or to conceive the intervening space bridged with something through which the effect is transmitted. When Clerk Maxwell in 1865 developed mathematically Faraday's ideas, it became clear that the propagation of electric forces was similar to the propagation of light ; whatever be the means by which one is transferred must also be the means of conveyance of the other.

Hence the idea of a universal all-pervading æther, which had been accepted owing to the success of the wave theory of light, received a great access of credit, and science entered on what might be described as an æthereal stage. Experiments were made to measure the velocity of the earth " relatively to the æther," and, when it appeared that they were relatively at rest, it was assumed that the earth carried the æther with it in its path, and attempts were made to demonstrate that whirling steel discs dragged the æther with them—once more with a negative result. However, a slight change in the dimensions of bodies with change in direction of motion—a change too small to be otherwise detected—was shown to be enough to explain the discrepancy, and it was universally assumed that time would bring a complete scheme of physics, in which the æther, interpreted in mechanical terms, would form a means of unifying the whole range of phenomena — an expectation which, indeed, has since seemed almost on the point of realization.

Maxwell's theoretical results were verified and interpreted to wider circles by the experiments of

Heinrich Hertz, who in 1888 demonstrated directly the passage through space of electro-magnetic waves, and showed that their properties were identical with those of waves of light of very great wave-length. By the invention of improved methods of producing and detecting these waves by Lodge, Marconi and others, they have become available for practical use in wireless telegraphy.

The work of Maxwell and Hertz for a time shifted the point of interest in electrical science from the The Theory electric charges and currents to the of Ions. dielectric medium. But, in more recent years, another line of investigation, simultaneously pursued, has concentrated the thoughts of men of science once more on the electric charge as a fundamental physical entity.

As we have already seen, the passage of electric currents through solutions of salts and acids produces chemical changes at the points where the current enters and leaves the liquid, the salts being there decomposed chemically into their constituent parts. To explain these phenomena, and to co-ordinate them with the other electrical properties of solutions, it is necessary to suppose that the opposite parts of the salt move in opposite directions under the action of the electric forces. Hence Faraday called these moving particles the " ions "—the travellers—and his conceptions have been successfully developed to explain the passage of the electric current as effected by the conveyance of discrete charges of electricity by moving atoms or groups of atoms of the dissolved salt. The motion of the ions may even be made visible to the eye by watching coloured ions move through an otherwise colourless liquid, and the velocities calculated by theory have been thus verified by experiment.

The ionic theory, framed to explain the conduction of electricity through liquids, has been adapted also to the case of gases. But the development of this subject belongs to a later period, and its consideration must, for the present, be postponed.

We have now reached what may be taken as the end of the period in physical science most characteristic of the nineteenth century. With the unchanging, apparently eternal chemical atoms, which "bore all the stamp of manufactured articles," as basis, science had built up a scheme which pictured a Universe of mechanism all but understood, and actuated by a limited amount of energy which could neither be increased nor diminished. Even the atoms were waiting expectantly for someone to find a successful hypothesis which would explain them away and turn them into contortions of the all-embracing æther. It seemed as though the main outlines of physical knowledge had been laid down once for all, and that the task of succeeding generations would be but to fill in those outlines, and carry to yet greater accuracy the measurements of known physical constants.

Summary.

At a later stage of our enquiry we must trace the revolution in outlook which soon disturbed this prematurely completed scheme, and the immense broadening and deepening of the range of known physical phenomena. For the present we must return and describe how a system of thought, the complement of that which had come to dominate physics, spread to the domain of biology, and was accepted as an adequate explanation of some of the deepest problems of life.

CHAPTER VI

THE COMING OF EVOLUTION

The Significance of Biology—Embryology—Natural History—Scientific Exploration — Organic Chemistry — The Physics of Physiology—Microbes and Disease—Geology—Evolution before Darwin—Darwin —The Fight for Evolution—The Effects of Evolution.

In the epoch of science which began with the Renaissance, it was astronomy that produced the greatest
The Significance of Biology.
revolution in the thoughts of mankind. It so happened that when Copernicus dethroned the earth from its proud position as the centre of the Universe, and Newton brought the phenomena of the heavens under the sway of the mechanical laws familiar in everyday life, they also undermined the evidence on which a whole theory of Divine revelation had been built up. A complete change of outlook was thus brought about, although many years passed by before the full effects were realized, and, as we must now show, the current conceptions were attacked in a fresh direction before it was possible to begin the work of reconstruction.

In the great advance which marked the nineteenth century, it was not the vast development of physical knowledge, and still less the enormous superstructure of industry raised on that knowledge as a basis, which most effectually widened man's mental horizon and led to yet one more revolution in his ways of thought. The point of real interest shifted from astronomy to geology, and from physics to biology and the phenomena of life. "Life," wrote Robert Hunter, the great surgeon of the eighteenth century, "is a property we do not understand. We can only see the necessary steps leading to it." It was this property that men now set themselves to examine critically. The necessary steps began gradually to show signs of falling into order in a scheme of natural knowledge, and the theory of natural selection, which gave an acceptable basis for the old and somewhat discredited idea of evolution, carried the human mind over the next long stage of its endless journey. Darwin was the Newton of biology—the central figure of nineteenth-century thought.

To trace the history and appreciate the significance of evolutionary philosophy, we must follow the outlines of the growth in biological knowledge from the point where we dropped the threads of its story in a previous chapter.

One of the principal lines of evidence in favour of the theory of evolution has come from a comparison of animals, a branch of enquiry which in the eighteenth century was in the hands of travellers and explorers; but there is another source of information which has proved no less convincing to the trained scientific mind. The formative process through which every living thing, animal or vegetable, passes before it takes up its final and

Embryology.

specific character has been shown to contain some record of the much more general and infinitely more prolonged development of the whole species.

Harvey had put the science of observational embryology on a correct basis in his *De Generatione Animalium* in 1651, but the true founder of the modern development was Caspar Frederick Wolff (1733–1794), who was born in Berlin, and died at St Petersburg, whither he had been summoned by the Empress Catherine and where he had long occupied a professorship. Wolff's work was neglected and discredited during his lifetime, but in truth he foreshadowed all the modern theories of structure. He made a study of cells, by means of the microscope, and showed the progressive formation and differentiation of the various organs in a germ originally homogeneous in character, thus destroying the previous belief that every organ made a separate and distinct start.

It is now known that the multiplication and differentiation of cells is a process common to all embryonic development, and that organic growth proceeds on identical lines throughout the whole animal creation. It becomes probable that the development of the higher animals repeats, within certain limits, the different stages of the process gone through by creatures who are lower in the scale of existence. It remained only for Von Baer, about 1820, to show that the pre-natal growth of man took place on lines identical with those of all other animals, and thus to forge one link in the chain of evidence which binds man to the rest of creation. It is perhaps difficult for the untrained mind to appreciate fully the value we must attach to such discoveries, which provide the most convincing and suggestive evidence to those who are accustomed to labour in these directions. Yet embryology disclosed, in the history of each indivdual, facts which were only to be synthesized with infinite difficulty and after much opposition, from the whole survey of animals, existing and extinct, as mapped out by the students of natural history.

After Buffon had accomplished his great work of description of the animal world, another Frenchman Natural took up the subject of classification, History. which always appeals to the Gallic mind, and placed it on a clear and definite basis. Georges Cuvier (1769–1832) was the son of a Protestant officer, who had migrated from the Jura district into the region of the Wurtemberg protectorate in consequence of religious persecutions. He spent the period of the Revolution and the Reign of Terror studying peacefully in Normandy, and, at their close, returned to Paris, where he soon occupied a foremost position in the Collège de France. His great claim to distinction is due to the fact that, first among naturalists, he compared systematically the structure of existing mammals, fishes, and molluscs with that shown by the remains of extinct fossil vertebrates and shells, so that the past was taken into account, no less than the present. Cuvier stands on the threshold of the new age of scientific discovery, and his great book, *Règne animal distribué d'après son organisation*, forms a connecting link between the work of men who studied the world and its phenomena as a stationary problem and those who were impelled to see in it a transitory stage of the great drama of evolution.

During the second half of the eighteenth century, the systematic exploration of the world proceeded Scientific apace, and much of it was undertaken Exploration. in a true scientific spirit. Captain James Cook (1728–1779), born at Marton in Yorkshire, and present at the capture of Quebec in 1759, had published work on a solar eclipse in the *Philosophical Transactions* of the Royal Society, before he sailed in the *Endeavour* to observe the transit of Venus from the South Pacific Ocean. His later voyages of discovery were conducted with keen appreciation of their opportunities for the advancement of knowledge, and he was one of the earliest recipients of the Copley gold medal of the Royal Society.

In any account of the progress of knowledge during the period under review, it is not easy to classify the work of Baron von Humboldt (1769–1859), the distinguished Prussian naturalist and traveller, who for a great period of his life found his most congenial home in Paris. He spent five years in exploring the continent of South America and the seas and islands of the Mexican Gulf. It was on observations collected during this expedition that he laid the foundations of the claims of physical geography and meteorology to be considered as accurate sciences. Von Humboldt was the first to map the earth's surface in lines of average equal temperature—isothermal lines—by which he obtained a method of comparing the climates of different countries. He studied the rate of decrease of temperature with increase of height above the sea-level, during his ascents of Chimborazo and other peaks of the Andes. He considered the origin of tropical storms and atmospheric disturbances ; he noticed the position of zones of volcanic activity, and suggested that they corresponded to subterranean cracks in the earth's surface ; he investigated the distribution of plants and animals as affected by physical conditions ; he studied the variations of intensity of the earth's magnetic force from the poles to the equator, and invented the term " magnetic storm " to describe a phenomenon first placed on record through his labours.

When he settled in Paris in 1808, to put into form and to publish the results of his explorations, he probably shared with Napoleon Bonaparte the distinction of being thought the foremost man in Europe. His *Kosmos*, containing the entire fruits of his life's work, was not produced till the end of his career, between the years 1850 and 1859. In these volumes von Humboldt endeavoured to give at once an accurate description and an imaginative conception of the world as known to our senses. It is impossible to attain complete success in an undertaking of such infinite magnitude and daring inception, and doubtless its abiding value will consist in the fact that it places on record the aspirations and comprehensive intelligence of a great man.

The interest excited by von Humboldt's labours and personality gave an impetus to scientific exploration among the nations of Europe. In 1831 the *Beagle* was despatched by Great Britain on her memorable expedition " to complete the survey of Patagonia and Tierra del Fuego ; to survey the shores of Chili, Peru, and some islands in the Pacific ; and to carry a chain of chronological measurements round the world."

The expedition was declared to be "entirely for scientific purposes," and Charles Darwin sailed on board as official " Naturalist."

A few years later (1839), Joseph Hooker (1817–1911), son of the well-known botanist Sir W. J. Hooker, joined Sir James Ross's Antarctic expedition and spent three years studying plant life amid the frozen seas ; proceeding later on to the northern frontiers of India, on an expedition of which the cost was also partly defrayed by the Government.

In 1846, T. H. Huxley left England as surgeon in the *Rattlesnake*, and spent several years surveying and charting in Australian waters ; his eager mind and keen powers of observation chafing at the lack of opportunity given for accurate scientific observations of general interest.

Thus three of the men who played a chief part in revolutionizing the thought of the nineteenth century, each served an apprenticeship on one of the scientifically planned voyages of exploration.

The culminating point of organized discovery and research on the grand scale was reached in the despatch of the *Challenger* in 1872, to cruise for several years in the waters of the Atlantic and Pacific, and to take a long series of records dealing with every branch of oceanography, meteorology, and natural history that came under the notice of the explorers.

The chemistry of the complicated substances which are found in the bodies of plants and animals is chiefly Organic Chemistry. the chemistry of the remarkable element carbon. The atoms of carbon possess the unique property of combining with each other, as well as with atoms of other elements, to form very complex molecules. We have seen how Paracelsus and Stahl carried on the old theory of a distinct vital principle in opposition to the equally old idea that in living bodies, as in the outside material world, mechanism would ultimately explain all happenings. It was long thought that the specially complicated substances which are characteristic of animal and vegetable tissue could be prepared only under the influence of vital processes, and the belief in a spiritual interpretation of the phenomenon of life was thought to stand or fall with the truth of this view. But the artificial preparation of urea by Friedrich Wöhler in 1828 showed that the methods of the laboratory were capable of creating substances hitherto reserved for living processes. Other preparations followed, till in 1887 Fischer and Tafel succeeded in making sugar from its elements. The secrets of the chemistry of the body seemed in a fair way to become known.

Simultaneously with the extension of chemistry to cover many vital changes, much advance was made The Physics in applying physical principles to the of Physiology. problems of physiology. Harvey explained the motion of the blood as that of a moving fluid pumped through the tubes of arteries and veins by the mechanical action of the heart, and thus gave a naturalistic turn to physiological enquiry. But, in the second half of the eighteenth century, the difficulty of the problem led to the almost universal adoption of the hypothesis of vitalism, the *force hypermécanique* of the French school, which held sway till the middle of the nineteenth century. Then the change of view begun by the synthesis of urea was reinforced on the physical side by the application by Mayer and Helmholtz to the living organism of the principle of the conservation of energy. All the manifold activities of the body were thus traced ultimately to the chemical and thermal energy of the food taken in, and it was natural, if not strictly logical, to conclude that, if the total output of energy was governed by physical laws, the intermediate processes could be described completely by those laws also.

The naturalistic standpoint was further strengthened by the work of Schleiden and of Schwann, who showed that all organisms were built up of living cells as biological units, and by other investigations on similar lines into cell-structure and cell-function. The knowledge of physical phenomena associated with solution, and especially with the solution of jelly-like substances or colloids, was rapidly applied to physiological problems, while the phenomena of nerve were found to have electrical concomitants. It came to be believed that physiology was but a special case of " the physics of colloids and the chemistry of the proteids." And, whatever be the truth about the whole physiological problem, and the psychological and metaphysical questions that underlie it, it is necessary for the increase of knowledge to assume that, at each step, the process is understandable by those natural principles which have already been elucidated, and find their best ultimate statement in the fundamental concepts of physics and chemistry.

One of the most striking developments of nineteenth-century science, and one which, by increasing our Microbes and direct control of the environment, has Disease. had a marked influence on our ideas of the relative positions of man and of "nature," was the growth in knowledge of the origin and causes of microbic disease in plants, in animals and in mankind. About 1838 Schwann discovered that the yeast present in fermentation consists of minute living vegetable cells, and that the chemical changes which go on in the fermenting liquor are due in some way to the action of the life of these cells. Schwann also perceived that putrefaction was a similar process, and showed that it too did not occur if precautions were taken to destroy by heat all existing living cells in contact with the substance examined, and to preserve it thereafter from contact with all air save what had passed through red-hot tubes. Thus both fermentation and putrefaction were proved to be due to the action of living micro-organisms.

These results were extended about 1855 by Louis Pasteur (1822–1895), who disproved the old idea of spontaneous generation in any known case, and showed that the presence of bacteria could always be traced to the entrance of germs from outside, or the growth of those already present. Pasteur thus showed that certain diseases, such as chicken-cholera and the silk-worm disease, were caused by specific microbes ; and gradually the germs characteristic of other diseases, many of them prevalent among mankind, have been discovered and their life-history traced. Though chronologically much of the results of Pasteur's discoveries have only been reached in recent years, it is well to trace the whole story at this point.

It has been found that, in some cases at all events, it is the presence of a definite substance in the microbic cells, or its production by their activity, that causes the changes associated with their life.   Thus in 1897 Büchner discovered the first-known of these active substances or enzymes in yeast cells, and showed that, when extracted from them, it could cause the same fermentation that living yeast cells produce.   Apparently the enzyme itself remains unchanged at the end of the operation ; its mere presence is sufficient to start and maintain the chemical action.

At the end of the eighteenth century, Jenner had introduced the process of vaccination, by which inoculation with the virus of small-pox, after its attenuation into cow-pox by transmission through a calf, causes partial or complete immunity to the severer form of the disease.   This principle of attenuation was extended to other diseases by Pasteur, and, in the case of rabies or hydrophobia, it was shown that inoculation was generally effectual even after infection. The mortality from this horrible and previously incurable disease was thus reduced to about one per cent. of the cases treated ; while in England the rigid enforcement of a muzzling order for dogs for a short period, and subsequent quarantine precautions against the reintroduction of the specific poison from the Continent, has freed the country from the shadow of this dread scourge.

The life-history of these pathogenic organisms is often very complicated, and some of them pass certain stages of their career in different hosts.   Only by most careful experiments, carried out by the inoculation of living animals, can their properties be investigated.   The hosts themselves sometimes remain unaffected by the invading microbe, and this immunity makes it exceedingly difficult to forecast the direction in which the source of infection must be sought.   The final conquest of malaria or ague is an excellent example of the difficulties and dangers which surround research in microbic disease.   Malaria, as the name shows, had always been associated with the bad air of swampy districts and undrained tracts of country. Many ingenious theories had been constructed and experiments undertaken, in order to show the connection between the poisonous gaseous exhalations of the marshes and the fever that they were assumed to produce.   The most convincing proof of the connection lay in the fact that those men who undertook to study the disease in its favourite localities invariably fell victims to the infection.   At last, the possibility of a secondary dependence only on the swampy nature of the district occurred to some of the investigators, and volunteers were found ready to spend the most unhealthy season of the year among the Italian marshes, having first provided themselves with wire and gauze protections against the insect life of the district. The experiment was successful: it was clear that the marsh alone was not the primary cause of the disease. The germ of malaria was discovered by Laveran, a French army surgeon, about 1880.   Five years later, Italian observers had shown that infection reached man from the bites of mosquitoes, and in the years 1894-7 Manson and Ross proved that one special kind of mosquito (*Anopheles*) was infested by parasites that turned out to be malaria germs in an early stage of development. Thus the correct method of attacking the ravages of malarial fever is neither to search for new drugs nor to hunt down existing mosquitoes, but to drain the marshes and to take note of all pools of standing water which provide suitable breeding-places, in order to destroy the larva by a thin coating of oil or other deterrent.

Similarly Maltese or Mediterranean fever has been traced to the action of a microbe which passes part of its life in goats, and is communicated to man by means of the goats' milk, the goats themselves remaining meanwhile apparently perfectly healthy. The connection between bubonic plague, rats, and the fleas or other parasites which help to bring the infection from the rats to human beings gives another instance of the indirect methods by which disease can now be fought with greatest success. By taking precautions founded on such new investigations, vast tracts of the earth's surface formerly uninhabitable, or only habitable at great cost of life and health, have been made fit for occupation. Of late years our general knowledge of microbic diseases, knowledge won by laborious and in some cases dangerous experiments, has led to ceaseless improvements in sanitation and preventive medicine on the one hand, and on the other, by the application of Pasteur's results by Lord Lister to surgery, has, in conjunction with the discovery of anæsthetics, made possible the wonderful operations now carried out to relieve human suffering. The practical results of these discoveries in hygiene, medicine and surgery are perhaps most clearly shown in the reduction of the annual death-rate of cities like London from about 80 per thousand two centuries ago to the present figure of some 15 per thousand. From the scientific aspect, the great alteration has been the opening up of a whole new world of organic life, previously undreamed of, existing in close contact with, and indeed thriving upon, the animal world previously believed to represent the sum total of life on this globe. The principal features of this discovery were the minute size of the organisms, by which they have hitherto escaped notice, their incalculable numbers, and the rapidity and simplicity of their reproductive methods, which enable them, despite their invisibility, to become, when they are of harmful nature, the most formidable enemies of the human race.

By his attempt to give a rational theory of the origin of the solar system, Laplace directed men's attention to the existence of the problem, Geology. and thus led to a quickened interest in the study of the earth itself—a study which had already begun to make some headway. Unfortunately, the countries in which freedom of thought had most prevailed against the claims of papal authority were also those in which the tyranny of the theory of the verbal inspiration of the Bible had been most firmly established, and a new contest, in which, however, the final issue was never doubtful, had therefore to be waged before any view of the earth's origin, other than that to be found in the book of Genesis, could win a general approval. Even in the early years of the nineteenth century, it was seriously contended that the fossils had been hidden in the ground by the Devil, the better to test the faith of mankind.

Some knowledge of rocks, metals and minerals had been acquired in the processes of mining at very early dates; but, although Leonardo da Vinci and Bernard Palissy in the fifteenth and sixteenth centuries had recognized in fossils the remains of animals and plants, they were generally regarded as *lusus naturæ*, and the products of a mysterious *vis plastica*, or a tendency in nature to produce certain favourite forms in various ways. The large collections of James Woodward (1665–1772), which form the basis of the geological collections of the University of Cambridge, did much to establish the view that fossils were really of animal and vegetable origin, but their use in telling the tale of the earth was not understood, save by scattered observers like Nicolaus Stensen (1669), whose ideas gained no general acceptance. Attempts made to explain how the earth reached its present state were still forced into accordance with the Biblical accounts of Creation and the Deluge, or were based on purely speculative cosmogonies involving cataclysmic origins by water or fire.

The first systematically to contend against these views was James Hutton (1726–1797), who published his *Theory of the Earth* in 1785. Once more a direct observation of nature and natural processes paved the way for scientific advance. Hutton, in order to improve the husbandry on his small paternal estate in Berwickshire, studied farming in Norfolk, and methods of agriculture in Holland, Belgium and Northern France. During fourteen years he pondered over the familiar ditches, pits and river beds, and then, returning to Edinburgh, laid the foundations of the modern science of geology. Hutton saw that processes able to produce stratified rocks and to embed fossils therein were still going on in sea, river and lake. " No powers," said Hutton, " are to be employed that are not natural to the globe, no action to be admitted of except those of which we know the principle "—a true precept of science, to avoid the framing of all unnecessary hypotheses.

Nevertheless, Hutton's " uniformitarian theory " was not generally accepted till William Smith had assigned relative ages to rocks by noting their fossilized contents ; till Georges Cuvier had reconstructed the extinct mammalia of the district from fossils and bones found in the Paris area ; till Jean Baptiste de Lamarck had made a classification and comparison of recent and fossil shells ; and finally till Sir Charles Lyell had collected all available evidence bearing on the manner and extent to which the earth is still being moulded into new forms by water, volcanoes and earthquakes, as well as all known facts about fossils, into his *Principles of Geology*, first published in 1830–3. The cumulative effect of long-continued processes was then fully grasped for the first time ; and the possibility was realized of tracing the history of the earth, at any rate throughout its habitable ages, from the record of the rocks, by inferences based on observation of operations that were still taking place.

The problem of the age of man is one of special interest to the human race. The discovery of flint implements, such as are still in use among primitive peoples, and of carved pieces of bone and ivory, in conjunction with the remains of animals that are now unknown or have long been extinct in Europe, enabled Lyell in 1863 to place man in a position in the long series of organic types, and to show that his existence on the earth must have extended over periods vastly greater than any contemplated by the accepted Biblical chronology. The subject is not yet closed. More recently, remains have come to light showing that man, as we know him now, was also preceded by types to which it is hard to give a name or to classify definitely either among the lower races of mankind or among the species of the higher anthropoid apes. The human form apparently has varied and developed, and is probably still subject to the possibility of change.

The idea of an evolutionary process in nature is at least as old as the days of the Greek philosophers. Heracleitus believed that all things were in a state of flux — πάντα ῥεῖ. Empedocles taught that the development of life was a gradual process, and that imperfect forms were slowly replaced by forms more perfect. By the time of Aristotle, speculation seems to have gone farther, and to have conceived the idea that the more perfect type might have developed out of, as well as preceded, the less perfect. But the atomists, who are often claimed as evolutionists, seem again to have contemplated the arising of each species *de novo*. Nevertheless, in their belief that only those types survived which were fitted for the environment, they touched in spirit the essence of the theory of natural selection, though their basis of fact was insufficient. But, in science, it has been truly observed that

Evolution before Darwin.

" being right is no excuse whatever for holding an opinion which has not been based on any adequate consideration of the facts involved in it." As in so many branches of knowledge, the time was not ripe for the Greeks to do more than formulate the problem, and make avowedly philosophic guesses at its solution.

Indeed, it needed two thousand years of time, and the labours of many quiet and unphilosophic physiologists and naturalists, before sufficient experimental and observational evidence could be collected to make the idea of evolution worth examining by science. It is a good illustration of the true scientific attitude of suspension of judgment in face of inconclusive evidence, that for the most part naturalists left the idea of evolution to the philosophers, and that, till Darwin and Wallace published their simultaneous work, the balance of scientific opinion was clearly against the theory. On the other hand, the philosophers also played their true part in maintaining speculation about a theory not ripe for scientific treatment, in keeping open a question of paramount importance, and in formulating solutions which, in due time, might serve as working hypotheses for the men of science with whom lay the ultimate decision. Hence it is in the nature of the case that, when, in the revival of learning, the idea of evolution once more appears, it is to be found chiefly in the writings of Bacon, Descartes, Leibniz and Kant, while the men of science were only slowly working far behind at facts which eventually would point in the same direction, through the embryology of Harvey and the system of classification of Ray. Moreover, some of these philosophers speculated on the possibility of quite modern methods in their conceptions of the present mutability of species and its experimental examination. On the other hand, it must not be overlooked that some philosophers who are claimed as evolutionists—forerunners of Darwin—took evolution in an ideal, not in a real, sense. Some of Goethe's views of evolution seem to have been of this nature, as were those of Schelling and Hegel. To them, the connection between species lay in the inner ideas which represented species in the conceptual sphere. " The metamorphosis," says Hegel, " can be ascribed only to the notion as such, because it alone is evolution . . . it is a clumsy idea . . . to regard the transformation from one natural form and sphere to a higher as an outward and actual production."

But this ideal outlook of some philosophers does not destroy the usefulness of the philosophic contribution to evolutionary theory. It is most interesting and remarkable that the division of labour and the difference of outlook between philosophers and naturalists was continued up to the very last moment. The philosopher Herbert Spencer was preaching a full-grown concrete evolutionist doctrine in the years that immediately preceded the publication of Darwin's *Origin of Species*, while as yet, from the want of definite evidence, the naturalists would have none of it. Philosophers were right, and naturalists were right ; they were each following the law of their being. The philosophers were dealing with a philosophic problem, one not ripe for scientific examination. The naturalists were exercising true scientific restraint in not taking, even as a working hypothesis, a speculation for which there was little observational or experimental evidence, and no satisfying suggestion of a mode of operation.

Nevertheless, through the eighteenth century and increasingly in the first half of the nineteenth, one naturalist after another ran counter to the prevailing consensus of scientific opinion, and upheld some form of evolutionary theory. First Buffon, who oscillated between the orthodoxy of the Sorbonne and a belief in " l'enchainement des êtres," put forward a theory of the direct modification of animals by external conditions ; next Erasmus Darwin, poet, naturalist and philosopher, caught a glimpse of the revelation which was to be given in its fullness to his grandson, and taught that, from " the metamorphoses of animals, as from the tadpole to the frog, . . . the changes produced by artificial cultivation, as in the breeds of horses, dogs, and sheep, . . . the changes produced by conditions of climate and season, . . . the essential unity of plan in all warm-blooded animals,—we are led to conclude that they have been alike produced from a similar living filament."

But the first connected and logical theory is that of Lamarck, who sought the cause of evolution in the cumulative inheritance of modification induced by the action of the environment. While, *pace* Buffon, the effect of change in environment is small on the structure of the individual, Lamarck held that, if the necessary changes in habits became constant and lasting, they would modify old organs, or, by the need for new organs, call them into being. Thus the ancestors of the giraffe acquired longer and longer necks by continually stretching for leaves on trees just beyond their reach, and the change of structure thus acquired became developed and intensified by inheritance. Though no direct evidence of such inheritance could be adduced, it gave a reasonable and consistent working hypothesis for future examination.

The attention drawn to the effect of environment on the individual, and the extent of the changes which may rightly be attributed to external circumstances, has probably had an enormous influence alike on thought and action. It is difficult to believe that, where the individual can sometimes be modified so profoundly, the species will remain unchanged. Hence, much of the social and philanthropic efforts of the nineteenth century were built up on the theory of modification through environment.

Two more evolutionists who maintained the direct action of the environment on the individual were Étienne Geoffroy Saint-Hilaire and Robert Chambers, whose anonymous book, *Vestiges of Creation*, had a great vogue, and helped to prepare men's minds for Darwin.

But the man to whom Darwin was solely indebted for the central idea of his work, the man who, strange to say, gave the same clue to Wallace also, was the Rev. Thomas Robert Malthus (1766–1834), at one time curate of Albury in Surrey. Malthus, an able economist, lived at the time when the expanding resources of the country, and its need for men to wage a war of existence, made possible and required a great growth of population. In 1798 Malthus published the first edition of the *Essay on Population*. In it he proclaimed that human population always tends to outrun its means of subsistence, and can only be kept within bounds by famine, pestilence or war, whereby the redundant individuals are eliminated. In later editions of the book, he admitted the importance of the prudential check, which then acted chiefly in the postponement of marriage, and thus, as far as man is concerned, destroyed his main contention in its striking simplicity.

Darwin has himself recorded the effect of this work on his mind. " In October 1838," he says, " I happened to read for amusement *Malthus on Population,* and being well prepared to appreciate the struggle for existence which everywhere goes on from long-continued observation of the habits of animals and plants, it at once struck me that under these circumstances favourable variations would tend to be preserved, and unfavourable ones to be destroyed. The result of this would be the formation of new species. Here then I had a theory by which to work."

The man to whom came this flash of insight was well fitted both by heredity and environment to make full use of it. Charles Robert Darwin (1809–1882) was the son of a remarkably able country doctor of ample means, Robert Waring Darwin of Shrewsbury. His grandfathers were Erasmus Darwin, of whom we have already written, and Josiah Wedgwood, the potter of Etruria, who also was a man of scientific power and ingenuity. The Wedgwoods were Staffordshire people, an old family of small landowners ; but the Darwins, of the same landed class, came from the Anglo-Danish country of Lincolnshire, and Charles Darwin's outward share of inheritance from the Northern race was made manifest by his tall stature and blue-grey eyes. Educated at Edin-

Darwin.

14

burgh with the idea of medicine, and at Christ's College, Cambridge, when he intended to take Holy Orders, Darwin got his best training as a naturalist on the five years' voyage of the *Beagle* in South American waters. In tropical and subtropical lands, teeming with life, Darwin received the impression of the interdependence of all living things, and within a year of his return he began to compile the first of his note-books on the facts bearing on the transmutation of species, led thereto by his study of South American fossils. Fifteen months later he read Malthus' book, and found the clue which enabled him to frame a theory of the means whereby new species might develop.

The individuals of a race differ from each other in innate qualities. If the pressure of existence or the competition for mates be great, any quality which is of use in the struggle for life or mate has " survival value," and gives its possessor an advantage which carries with it an improved chance of prolonging life, or of securing a mate and of rearing successfully a preponderating number of offspring to inherit the favourable variation. That particular quality therefore tends to spread throughout the race by the progressive elimination of those individuals who do not possess it. The race is modified, and a different and permanent variety may slowly be established. This was the new conception ; and it was well put by Thomas Henry Huxley, who by his power of exposition, skill in dialectic and courage in controversy did more than any other man to compel general acceptance for the views of Darwin and Wallace :—" The suggestion that new species may result from the selective action of external conditions upon the variations from the specific type which individuals present—and which we call 'spontaneous' because we are ignorant of their causation—is as wholly unknown to the historian of scientific ideas as it was to biological specialists before 1858. But that suggestion is the central idea of the *Origin of Species* and contains the quintessence of Darwinism."

With this idea as a working hypothesis, Darwin spent twenty years collecting facts and making experiments. He read books of travel and treatises on sport, natural history, horticulture, and the breeding of domesticated animals. He carried out experiments on the crossing of tame pigeons ; he studied the transport of seeds, and the geological and geographical distribution of plants and animals. In the assimilation of facts, in appreciating their bearing on all the complicated questions which arose, and in marshalling them at the last, Darwin showed himself supreme. His transparent honesty, burning love of truth, and calm and even balance of mind, form a model of the ideal naturalist. Fertile in hypotheses as a guide to work, he never let a preconceived view blind him to facts. " I have steadily endeavoured," he writes, " to keep my mind free so as to give up any hypothesis, however much beloved (and I cannot resist forming one on every subject), as soon as facts are shown to be opposed to it."

By 1844 Darwin had convinced himself that species were not immutable, but still he worked on year after year to gain yet surer evidence. In 1856 Lyell urged him to publish the results of his researches ; Darwin, not satisfied with their completeness, delayed. On June

18th, 1858, he received from Alfred Russel Wallace a paper, written in Ternate in the space of three days after reading Malthus' book. In this paper Darwin at once recognized his own theory set forth in its essence. Unwilling to seize the priority of twenty years, which was rightly his but might destroy the interest of Wallace's contribution, Darwin placed himself in the hands of Lyell and Hooker, who arranged with the Linnæan Society to communicate on July 1st, 1858, Wallace's paper together with a letter from Darwin to Asa Gray dated 1857, and an abstract of his theory written in 1844.

Darwin then set to work and wrote out in condensed form the results of his labours of twenty years, and on November 24th, 1859, this book was published under the name of *The Origin of Species*.

We have already traced the various converging streams of thought—cosmological, anatomical, geo-logical and philosophic, which, blocked by The Fight for Evolution. the evidence in favour of the fixity of species, were yet collecting deeper and ever deeper behind it. Darwin's great contribution to know-ledge, driven forward by the quickening force of the conception of natural selection, broke the barrier with irresistible power, and set loose the fertilizing torrent over all the realms of science and of human thought.

At first the stream seemed to many people a devas-tating flood, obscuring without reason the whole out-look of the human race ; but we should not condemn without consideration the attitude of those to whom the new knowledge appeared impious and distasteful.

Now that the conception of evolution has become general property, it is difficult to realize how very limited were the numbers of men who were in a position really to judge of the value of the evidence laid before them, evidence that depended on the detailed examination of living creatures and fossil remains, forms unfamiliar and for the greater part unknown to those who felt compelled either to deny the validity of the conclusion or to give up beliefs which had sustained long generations of their forefathers. Before we condemn them, let us ask ourselves honestly whether, in ignorance of the intermediate stages, it is more obvious to postulate a common ancestor or a separate creation for the frog and the peacock, the salmon and the hummingbird, the elephant and the mouse; or, to go back to an earlier stage in the evolution of knowledge, let us make a list of the number of persons in the circle of our acquaintance who could immediately offer convincing evidence of the three-century-old heliocentric theory of the Universe.

And even to some naturalists the new ideas were repugnant. Owen, the great anatomist, wrote a strongly adverse criticism in the *Edinburgh Review*, and most systematists agreed with his opinion. But Hooker gave in his adhesion at once, and was immediately followed by Huxley, Asa Gray, Lubbock and W. B. Carpenter, while Lyell announced his conversion at the Royal Society Dinner in the autumn of 1864.

From the first, Huxley was the protagonist of this band of evolutionists—" Darwin's bulldog," as he called himself. With magnificent courage, ability, and clearness of exposition, he bore the chief brunt of the attack made from all sides on Darwin's book, and again and again led successful counter-attacks on his discomfited foes.

Thomas Henry Huxley was born at Ealing in 1825, the seventh son of an assistant schoolmaster ; but his parents came from families located at Coventry and on the Welsh Marches, and Huxley himself had the true fighting temperament of a border race. Huxley tells us that, to the men of science of that generation, the publication of the *Origin of Species* had the effect upon them of a flash of light in the darkness. " We wanted," he writes, " not to pin our faith to that or any other speculation, but to get hold of clear and definite conceptions which could be brought face to face with facts and have their validity tested. The *Origin* provided us with the working hypothesis we sought. Moreover, it did the immense service of freeing us for ever from the dilemma—Refuse to accept the creation hypothesis, and what have you to propose that can be accepted by any cautious reasoner ? In 1857 I had no answer ready, and I do not think that anyone else had. A year later we reproached ourselves with dulness for being perplexed with such an enquiry. My reflection, when I first made myself master of the central idea of the *Origin*, was, ' How extremely stupid not to have thought of that ! ' "

The famous scene between Bishop Wilberforce and Huxley at the Oxford meeting of the British Association in 1860 has often been described. Wilberforce had obtained a first-class in the Oxford Mathematical Schools in his youth, and therefore, being regarded by his University as a master of all branches of natural knowledge, had been selected to uphold the cause of orthodoxy. The Bishop endeavoured to kill the

notion of evolution with ridicule and sarcasm—
ridicule for Darwin and his labours, sarcasm for Huxley
and his courage.   It seems strange now to think that
a majority of the hearers were probably on the side of
the Bishop, and were totally unable, from preconceived
ideas, to weigh the value of the facts laid before them
on behalf of Darwin's theory, or to appreciate the
embryological evidence for evolution on which Sir
John Lubbock, now Lord Avebury, insisted.

It would be a mistake to think that the evidence
offered in favour of some form of evolution through
natural selection was considered conclusive, even by
Huxley himself.   Two years after the Oxford meeting,
Huxley wrote to Sir Charles Lyell :—

" If Darwin is right about natural selection—the
discovery of this *vera causa* sets him to my mind in a
different region altogether from all his predecessors—
I should no more call his doctrine a modification of
Lamarck's than I should call the Newtonian theory
of the celestial motions a modification of the Ptolemaic
system.   Ptolemy imagined a mode of explaining those
motions.   Newton proved their necessity from the
laws and a force demonstrably in operation.   If he
is only right Darwin will, I think, take his place with
such men as Harvey, and even if he is wrong his
sobriety and accuracy of thought will put him on a
far different level from Lamarck."

At one point in the controversy, much had been
held to depend on certain supposed differences of
structure in the human brain as compared with that
of the anthropoid apes.   Huxley set to work to throw
light on the matter, " skull-measuring all day at the
College of Surgeons."   As the result of this study he

extended his field of observation, and helped to create the modern science of ethnology, of which we shall have occasion to speak on a later page. Of Huxley's work in this direction, Professor Virchow, who led the Continental ethnologists, said in 1898 that it was sufficient to secure immortal reverence for his name.

In attempting to estimate the influence of the establishment of the theory of evolution on philo-

The Effects of Evolution. sophic thought and on the general opinions of mankind, we are faced by the difficulty that we are still in the period which is under the domination of evolutionary ideas as a new force in the realm of knowledge. We are still testing evolution, natural selection and all the kindred conceptions they have called into being as guides for our footsteps in science, philosophy, sociology and religion. Hitherto, the main concepts have stood the ordeal well, and the modifications which experience has suggested have been in matters of detail. But we are still too near the date of the publication of the *Origin of Species* to see the whole of its consequences in true and just proportion. Nevertheless, certain conclusions stand out clearly, and may be stated with some approach to confidence.

As the history of thought moves on, the mechanical and spiritual theories of the Universe alternate with each other in recurring pulsations which are as necessary to a healthy growth of knowledge as the alternate diastole and systole of the heart are to the well-being of the body. With each great advance in scientific knowledge, with each subjection of a new kingdom to the rule of natural law (as the process comes to be

regarded), the human mind, by an inevitable exaggeration of the power of the new method, tends to think
that it is on the point of reaching a complete mechanical
explanation of the Universe.   The Greek atomists
made a guess at the structure of matter which chances
to concord with modern views, though their evidence
for it from the scientific point of view was most
exiguous.   Not content with applying their theory
to the inorganic world, they framed accounts and
explanations of life and its phenomena on the idea
of a " fortuitous concourse of atoms," all unconscious
of the vast complexity of inorganic nature, and the
still vaster world of new phenomena which had to be
explored before the problem of life, for which they
gave a confident solution, could even be approached.
Yet the atomists did good work, and did it under
the inspiration of a wave of materialistic philosophy.
But the insufficiency of their conceptions was recognized by Plato and Aristotle, and, for equally transient
reasons, an idealistic spiritualism returned.

The faith of the Middle Ages, where it depended
on the exercise of the human reason, was constructed
from argument based on doubtful premisses, and
fortified by a theory of divine providence adapted
to an undeveloped science.   Hence, when the growth
of knowledge began afresh with the Renaissance, the
natural oscillations of opinion were increased.   Each
step in advance was feared as an offence against
orthodoxy, and its leader was denounced as an
impious heretic.

But here again we must not condemn the conserva-
tors of ancient creeds without some thought.   They
had inherited a scheme of religion and philosophy,

complete in itself, which satisfied the majority of their followers, and was in accord with nature, as appearing in their daily life. The new facts brought to light in the upheaval of the Renaissance, although they put out of gear several minor parts of the great mediæval scheme of salvation, were not sufficient of themselves to justify or to permit the reconstruction of a whole new consistent model of theological opinion and religious practice. The mass of mankind require emphatic warrant for their beliefs, and cannot be reconciled to the twilight of faith in which greater minds can soberly acquiesce. It may be perfectly true that there is sometimes more faith in honest doubt than in the majority of oft-repeated creeds; but this is not a proposition of general application, as the great Lincolnshire poet and mystic—another typical Anglo-Dane—well knew. Therefore those people may have been well advised who tried blindly to maintain the ancient beliefs until a time came when it should be possible to put something more satisfying than mere negation in their place.

The triumph of Copernicus, the success of Newton in interpreting the phenomena of the heavens, led up to the overestimate of the power of their methods by the French Encyclopædists, who at once conceived a mind competent to calculate the whole of the future history of the world from a knowledge of the initial configuration and velocities of the masses. At each step made, this joyful overestimate of the possibilities of mechanism becomes a marked feature of contemporary thought. As each piece of knowledge becomes assimilated, it is seen that old problems are in their essence unaltered; the poet, the seer and the mystic again come to their own, and, in new language, and from a higher ground of vantage, proclaim their message to mankind.

Now, speaking broadly, this recurrent phenomenon of a wave of mechanical philosophy was the first main result of Darwin's success. Quite legitimately and without exaggeration, the establishment of the principle of evolution strengthened greatly the feeling of continuity in nature, and gave new confidence to those who based their view of life on scientific ground. It was the complement on the biological side of the contemporary tendencies in physics we have traced in the last chapter—tendencies which pointed to a complete account of the inorganic world in terms of eternal, unchanging matter, and a limited and strictly constant amount of energy. These conceptions seemed to destroy the need for an overseeing Providence or for any act of creation, or at the least implied that the Author might well be supposed to have turned away, and left the great machine to spin on unheeded, unwatched, down the ringing grooves of change. The application to living beings of the principle of the conservation of energy, led to the exaggerated belief that all the various activities, physical, biological and psychological, of the existing organism would soon be explained as mere modes of motion of molecules, and manifestations of mechanical or chemical energy. The acceptance of the theory of evolution produced the illusion that an insight into the method by which the result had been obtained had given a complete solution of the problem, and that a knowledge of his origin and history had laid bare the construction of the inward spirit of man as

well as of the human organism regarded from without.
Now whether a complete knowledge even of the outward problem, based on mechanical principles, will
ever be reached it is impossible to say. But it is
quite certain that as yet it has not been attained, and
will not be attained till many more alternations
towards and away from mechanical philosophy have
passed like waves over the human mind. Indeed, we
are already moving out of the particular wave induced
by the coalescence of nineteenth-century physics and
the coming of evolution. The present tendency is on
the whole away from thorough-going materialistic or
even mechanical views. Our opinions as to whether
this tendency is in the direction of final truth, or
whether it is but a passing eddy on the broad current
of thought which, with temporary set-backs, moves
for ever onwards, will depend on the inborn qualities
and predispositions of our minds, as modified by
education and environment. Moreover, it must be
remembered that the type of mind to be satisfied
may alter. The mystic of the Northern race will
never rest content with a mechanical universe if it
be held to regulate inward consciousness as well as
outward phenomena. Minds which rely on law and
form will require some ordered and comprehensive
scheme to satisfy their mental outlook. But whatever
be the outcome, it must be admitted that the problem is too vast to be solved in any finite time. The
very principle of evolution itself requires us to look
forward to an ever-changing stream of thought, which
will develop from age to age, while past experience
goes to show that the development will be not steady
and secure, but intermittent and oscillatory.

In one sense, the acceptance of the theory of natural selection is the completion of the philosophic work begun and mapped out by Francis Bacon. Bacon taught that the method of empirical experiment was the sole road to natural knowledge for mankind.   Darwin showed that Nature herself uses the method of empirical experiment, both in the animal and vegetable worlds.   She tries all possible variations, and, out of countless trials, a few succeed in establishing that new and greater harmony between the being and its environment from which evolution proceeds.

The Experi- ments of Nature.

If accepted in its fullest sense, natural selection is the negation of all teleology.   There is no end in view : merely a constant haphazard change both of individuals and of environment, and sometimes a chance agreement between them, which, for a brief moment, may give some appearance of finality.

Herbert Spencer's phrase for natural selection, " the survival of the fittest," standing alone, begs the question.   What is the fittest ?   The answer is : " The fittest is that which best fits the existing environment."   It may be a higher type than that which preceded it, but it may be a lower.   Evolution by natural selection may lead to advancement, but it may lead to degeneration.   As Mr Balfour has pointed out, on the full selectivist philosophy the only proof of fitness is survival :—that which is fit survives, and that which survives is fit.   We may seek to break away from the circle by declaring that, on the whole, evolution has produced a rise in type, that man is higher than his simian ancestors.   But then we take on us to pronounce on what is higher and what is lower, and the thoroughgoing selectionist may reply that our judgment is itself formed by natural selection, and thereby is framed to appreciate and rate as higher that which, in reality, merely has survival value— that which, in fact, has permitted us to exist.   From the purely naturalistic standpoint, there seems no escape.   We have to accept an absolute judgment by some other standard of what is high and low, good and evil, if we seek another outlook.

Indeed, it may be pointed out that the order in which we place creation is largely a matter of race and racial religion.   To the oriental Buddhist, existence is an evil, consciousness a greater evil.   To him, then, logically, the highest form of life is a simple cell of protoplasm in the tranquil depths of the ocean's bed, and all the evolution of the ages is in truth downward from that calm ideal, which is itself a fall from the inorganic matter that probably preceded it.

Of all the sciences regenerated by Darwin, none gained more benefit than anthropology, the comparative study of mankind.   Indeed, it Anthropology. is hardly too much to say that modern anthropology took its rise from the *Origin of Species.* Huxley's study of human skulls was the beginning of that exact measurement of physical characters on which so much of the science now depends, and the ideas of natural selection and evolution underlie all succeeding work.

Yet the ground had been prepared for anthropology also.   The same love of novelty, the same eager curiosity, the same acquisitive collector's instinct, which introduced the plants and animals of other climes into the European gardens and museums, brought back the artistic and industrial products of other people and the ceremonial objects of other religions, in all stages of development. First as a source of wonder and interest, then as a matter for commiseration and missionary enterprise, and finally as objects of admiration, study, and investigation, we of the West have come into contact with older civilizations than our own—older in origins, older and far anterior to our own in the sequence of the ages.

The ingenious diplomatist who enriched our British Museum with the sculptures of the Parthenon, the men who first brought back from Asia Minor and Babylonia the spoils of earlier ancient empires, were not anthropologists in our present sense of the word. They were true treasure-seekers—pirates and buccaneers of the Northern art world—at best, bent on confirming, for their own satisfaction, the most striking passages of the classical historians who ruled in the schools of the period. The plunder of their Eastern voyages was deposited in the capitals of Western Europe ; trophies of the chase, which have taught us that art is but one aspect of the development of the human mind—as a record, incomplete in itself—and that history is but the detailed account of human evolution, as its progress appears to the subjects of the experiment. Thus, when the anthropologists were ready to take the field, much of the necessary material was at hand, already familiar, or partially classified, awaiting only the new gift of reinterpretation to give up another aspect of its inward meaning.

As an example of modern methods and results we may take the analysis of the population of Europe, to which reference has already been made. Considered physically, European people differ chiefly in three characters :—stature, coloration, and skull-shape. On the average of large numbers, moving northward, until we approach the arctic region, the stature becomes greater and the colouring fairer, while, the farther to the south we go, the shorter and darker becomes the population. In the intermediate Alpine region, stature and colouring are intermediate also. But the head shape tells a different tale. While both north and south folk are long-skulled, the people of the hills are broader skulled than either of the others.

To explain these facts we have to suppose three primary races—a tall fair race, found in its greatest purity round the shores of the Baltic, a short dark race stationed about the Mediterranean coasts and up the Atlantic seaboard, both with long skulls, and a broad-headed race, intermediate in colour and stature, inhabiting the mountainous regions of Central Europe. We have already pointed out that the history of Europe is the story of the interaction of these three races.

Similar methods of investigation, using often other characters such as the texture of the hair, have been applied to the physical anthropology of other continents, where more primitive folk are to be found.

The *Origin of Species* raised anew the great question of the origin of the human race. It was treated in detail in 1863 by Huxley in his book, *Man's Place in Nature*, and by Darwin himself in 1871 in a work named *The Descent of Man*.

It is a curious psychological fact that, whether in studying family history or in speculating about the origin of the human race, man prefers to believe that he has fallen from the state of ancestors better than himself rather than that he has risen in the social or racial scale. This touching faith in the efficacy of heredity as opposed to the merits of education, environment, and personal endeavour appears to be ingrained in our constitutions. Like other such prepossessions, it has probably some survival value, and should be treated with more respect than it often encounters. Hence new men may be forgiven when they provide themselves, if Nature and the College of Arms have omitted to do so, with noble forebears, and primitive races when they postulate a direct descent from the gods, or, if somewhat more modest, ask only a specific divine act of creation to have been exercised in their favour. And civilised man, confronted with the choice between the Book of Genesis and the *Origin of Species*, at first naturally proclaimed with Disraeli that he was " on the side of the angels."

Yet the evidence of man's affinity with animals was overwhelming, and soon prevailed within the limited circle where rational discussion was possible. His body contains the same tissues built into the same organs. He bears numerous vestiges of structures useless to him, but advantageous to his lowly ancestors. Many of his emotions and mental processes are evident in less developed states in the minds of animals. His origin is explicable by natural and sexual selection from hypothetical forms more ape-like than himself. His very weakness and defencelessness against the wild beasts, his earliest foes, may be held to explain his social instincts, the origin of moral qualities, for by them alone could he learn to combine for defence, and to develop speech and those mechanical arts whereby he obtained homes and tools and weapons.

Since the days of Darwin, discoveries have been made whereby the existence of modern races has been pushed back into more remote times than were formerly thought conceivable. We know that twenty-five thousand years ago the cave-men artists were decorating their walls with spirited likenesses of the bison and wild boar. At Neanderthal in 1856, and at Spy in 1886, still older remains have come to light, showing the existence of more primitive forms of man ; and in 1893 bones were discovered in Java which most authorities hold to be those of a being inter-mediate in structure between the anthropoid apes and the earliest known forms of men.

Whether man descends from any existing form of ape is uncertain. But if not a lineal descendant, he is at least a distant cousin. More variable forms may have preceded all those now extant, and have been their common ancestors. The separate shoots that rise above the visible plain of history may spring from a common root-system, buried in the dark ground of the irrecoverable past.

As Copernicus removed the earth from its position at the centre of the Universe, so Darwin took man from his cold pedestal of isolation as a fallen angel, and forced him to recognize his kith and kin in St Francis's little brothers, the birds. As Newton proved that terrestrial forces hold sway in the heights of heaven and the depths of space, so Darwin showed that the familiar variation and selection by which man moulds his flocks and herds may explain the development of species and the origin of man himself. Organic nature, like physical nature, became one problem. Creation, if never to be understood, at least is something to be investigated by methods we can understand—a new and mighty revelation to the human mind.

Evolutionary philosophy has modified profoundly our conceptions of human society. In a word, it has History, destroyed for ever the idea of finality. Politics, and Never again shall we set forth to build Sociology. cloud-cuckoo-boroughs between heaven and earth, nor sail across the seas to found a new Utopia with More, or a Pittsburg, Pa., with the Pilgrim Fathers. The legitimist Tory, with his unchanging, divinely ordered State, and the philosophic Radical with his ideal Republic only delayed by Tory stupidity and cupidity, become equally obsolete. They go, with the hansom cab in which they drove, to South Kensington Museum. Political institutions, no less than living beings, must fit their environment. Both are subject to variation, and, for social health, must develop *pari passu*. Institutions successful in one racial environment may fail lamentably in another. Even Liberals are coming to learn that representative government on the British model may not be applicable to every nation. The Hottentot may be right in preferring the clear evening fire that awaits him in his primeval kraal to tea on the terrace of the House of Commons at Westminster.

A similar change is passing over economics. The formal political economy of the earlier days of the science sought for a state of society, eternal, universal, valid for all times, in all places, for all peoples. The economist desired to emulate, without understanding, the discoveries of Newton. His view, analogous to the immutability of species of the older zoology, was undermined by the historical school, which, in many directions, has shown that economic laws are only valid in suitable circumstances, and that their application changes with the ever-varying environment.

But, as in biology, the changes in political institutions and in economic conditions are alike slow. We cannot take short cuts to the next stage by means of a surgical operation ; nor indeed do we know where the next stage will lead us. Survivals of past times are found side by side with rudimentary forms ready for new growth. As morphology discloses in the animal body vestiges of organs useful in past phases of organic evolution, so the study of social institutions shows traces of the older stages through which they have passed. From these traces, rightly interpreted, their history and origin may ofttimes be inferred. And from a knowledge of history and origin light is cast on present meaning and true significance. Such books as Tylor's *Primitive Culture* and Frazer's *Golden Bough* show how the concepts of evolution illumine the study of sociology.

If man has been brought into being by the same processes of evolution as the animal races, he must be subject still to the same variation and selection. Darwin's cousin, Francis Galton, working on this idea, traced the inheritance of physical and mental qualities in mankind, and showed that, to secure the continued progress of the race in the direction that civilized men have agreed to consider upward, even to prevent its deterioration, it was necessary that selection should continue.

In civilized states, it is probable that the most effective selective agent is disease. Those specially liable to any particular weakness tend to die early and leave no offspring. Thus, the hereditary predisposition to the complaint is bred out of the race.

But, in order that natural selection should be effective, it is necessary that successful survival should be accompanied by preponderating reproduction. From the point of view of the race, it is of no use for the higher types of men to survive if they do not leave children to inherit their qualities. Hence the power of voluntarily restricting the birthrate, which has become a marked feature of European civilization since 1876, has modified fundamentally the biological problems of sociology. The thrifty and far-seeing in all ranks of life now tend to restrict closely the number of their children. Their average family has sunk from over six to about three, and shows signs of further diminution. On the other hand, the careless and improvident, and, still more ominous, the mentally weak, still breed at nearly their old natural rate ; while, owing to the advance of modern scientific hygiene, they have a greatly diminished death-rate, and rear a high proportion of decadent citizens. The good hereditary qualities of the race are threatened with decay. It is probable that nothing short of the religious motive, which is certainly effective among many Jews and Roman Catholics, will suffice to reverse these tendencies of the better stocks. In this way, religion would acquire once more an effective survival value.

When we pass to the bearing of Darwin's work on theories of politics, no consensus of opinion has been reached.   The principle of the survival of the fittest has been used to revivify aristocratic ideas by Vacher de Lapouge, Bourget, Ammon, and Nietzsche.   But, on the other hand, it has been urged that evil qualities may have an advantage in present conditions ; that a secure aristocratic position removes competition and therefore selection ; that " equality of opportunity " is of the essence of Darwinian progress.   Moreover, socialists point to the societies formed by animals for mutual aid and their great " survival value," finding in the lives of the bee and the ant arguments for a communistic order of society, and possibly, also, for a finality of development, which might end by becoming stationary, could such a system of control become universal.   The bee world has shown no sign of progress during the two thousand years that it has been under observation.   It is rigid, utilitarian, self-sufficing—a model for communal life, when human desire and individual initiative have been bred out of the race.   In truth, the application of the principle of natural selection to sociology is so complex a problem that almost any school of thought can draw illustrations from an endless series of examples, choosing them at will, and may thus obtain valid arguments for their special tenets from the *Origin of Species.*

If the influence of the work of Darwin and his successors on sociology has been great, no less pro-
Religion.   found has been its effect on religious thought.   The destruction of the crude dogma of separate specific acts of creation, though the most immediately obvious, was but the most superficial of its results.

The principle of natural selection weakened immeasurably the old " argument from design " of the Christian apologists. Adaptation of means to ends in plants and animals received a naturalistic interpretation, which, if not complete in the deepest recesses of the problem, went far towards solving it. It was thought to be no longer necessary to invoke an intelligent and beneficent Artificer to explain the details of bodily structure or the protective markings on a butterfly's wing. The need for a Supreme Being, Jehovah, Jove, or Lord, was driven further back, inward, into the heart of man.

But the destruction of untenable positions was not the only service rendered to an unwilling theology by the revelation of evolution. Forced to reconsider its premises, it created the new spirit of reverent enquiry, of co-operative thought, of receptive attention. The Metaphysical Society was founded in 1869 by James Knowles, afterwards editor of the *Nineteenth Century*, with the encouragement of Alfred Tennyson. The meetings brought together such men as Archbishop Thompson, Bishop Ellicott and Dean Stanley, Cardinal Manning and W. S. Ward, W. E. Gladstone, Lord Selborne, Henry Sidgwick, James Martineau, Frederic Harrison, Lubbock, Huxley, Tyndall, Andrew Clark, John Morley and Leslie Stephen ; while Tennyson, Ruskin and many other distinguished men were present at the discussions. The personal relations between members of the society made wilful misunderstanding and ill-feeling impossible, and thus created an opportunity for distinguishing between the essential and unessential differences in the issues at stake, and for initiating a change of attitude in the leaders of religious thought in accordance with the sudden expansion in the domain of natural knowledge.   In place of the theory of a rigid and complete body of doctrine, delivered once for all to the Saints — a theory constantly liable to dislocation from the shocks of historical discovery— we gain the conception of an evolution of religious ideas, of revelation if we will, marked by supreme outpourings at certain times, but never ceasing to interpret the Will of God to mankind.   The gift of prophecy is restored to us.   We are driven back to that observational method in religion which has proved so necessary in science, to the study and recognition of the validity of individual religious experience, to the mystical standpoint, which once again proves a sure rock of defence.

Finally, we are led to the careful consideration of religious origins, to the reverent examination of those primitive religions through which the " heathen in his blindness " teaches civilized man the psychological meaning of his own ritual and his own creeds, and to the exploration of the unknown, transcendental and subliminal manifestation of the human mind — of which traces are to be found in nearly all societies, religions and races.   But the outcome of these researches belongs to a later phase of human thought, and will be dealt with in the next chapter.

CHAPTER VII

THE LATEST STAGE

THE last two chapters have described the characteristics of the science of the nineteenth century, and their
The New Dispensation. bearing on the eternal problems that face the human mind. We have seen how the physical concept of everlasting unchanging atoms, placed in a Universe containing a definite and limited amount of energy, was pointing anew to the old theory of a rigid determinism, described by mechanical laws and leading to a hard, materialistic philosophy. This tendency was increased and completed by the application of the principle of the conservation of energy to the functions of living beings, and by the theory of natural selection, which undermined theological ideas then current and gave to evolution a method of action independent of teleological influence.

We have now to trace the modification of the attitude of mind induced by nineteenth-century science. We must describe the widening of mental horizon caused by recent physical discoveries, and note the gradual recognition that, however successfully the mechanism of physiological processes may be described in physical or chemical terms, and however clearly the story of organic evolution may be pictured by natural selection, the fundamental problems which underlie the mysteries of consciousness and of the existence of the Universe remain unsolved if not unaffected.

Matter is more complex than was thought. The atom has been resolved into units of electricity, and shattered into a thousand fragments. Energy, though constant in amount, continually becomes less available in all physical conditions known to us; but those conditions may be far from exhausting the possibilities of creation. Darwin and Mendel have thrown light on the methods of evolution and the laws of inheritance, but the confident anticipations of some of Darwin's early followers that the wonders of creation were in a fair way to cease to be wonderful, have passed away. A naturalistic explanation of external Nature may be a little less remote, but the secret of the inner consciousness of man remains untouched.

Till the year 1895, it seemed that the main lines of physics, laid down during the middle of the nine-teenth century, had settled the limits, as well as the foundations, of the sciences of matter and energy. There seemed little left to do but to fill in the details of the scheme. But at that date a new series of phenomena began to come to light and soon revolutionized the whole outlook.

The Expansion of Physical Science.

The passage of an electric discharge through rarefied gases had been studied by Crookes, J. J. Thomson and others for many years, and the theory of electric ions, framed, as we have seen, to explain the conduction of electricity through liquids, had been adapted to this case also. But in 1895 W. Röntgen of Munich discovered accidentally that photographic plates standing in the neighbourhood of an electric discharge tube were affected though covered with an opaque screen, and thus revealed the rays associated with his name.

When an electric discharge is passed between two wires in a gas which has been very highly rarefied, Corpuscles straight rays are found to proceed from and Electrons. the negatively electrified wire or cathode, and to produce Röntgen rays when they strike solid objects. These straight cathode rays have been studied closely, especially by Sir Joseph John Thomson and his pupils. Thomson has also more recently examined the positive rays which, also in straight paths, proceed from the positive wire or anode. By making these rays pass through intense fields of magnetic and electric force, they may be deflected from their straight course, and, from the deflection caused by known forces, their velocity and their mass can be calculated. They are found to consist of flights of isolated atoms, and the experiments give an accurate means of measuring the atomic weight.

By similar methods, the negative or cathode rays had already been examined. They were shown to consist of flights of more minute particles negatively electrified, and, again by measuring the electric and magnetic forces required to deflect the particles from their straight path, Thomson was able to calculate their velocity and their mass. The velocity proves to range round a value of about 18,000 miles a second, or one-tenth that of light, while the mass of each of them is about the eight-hundredth part of the mass of the lightest chemical atom known—the atom of hydrogen. These " corpuscles " seem to be identical from whatever source they come, and whatever be the conditions of the experiment. They form a constituent part of all chemical elements, while the charge of electricity they carry seems to be the true natural unit of negative charge. They approach, at all events, the old conception of a common basis for matter formulated by Greek thinkers two thousand five hundred years ago.

But, at this point, the ideas we have been following come in contact with those suggested by another and independent enquiry, set on foot by the mathematicians Lorentz and Larmor. Maxwell's work had shown long ago that light was to be regarded as a series of electromagnetic waves. It followed that the waves must be started by the oscillations of minute electric charges, and that every substance capable of emitting light when raised to incandescence must contain electric charges ready to be set in oscillation. But moving electric charges carry electromagnetic energy and momentum with them in the surrounding space, and thus themselves possess inertia or mass. Now these two properties are sufficient to endow matter with its known characteristics, and thus matter itself may be explained as an electrical phenomenon, while the ultimate electric charges or " electrons " may be identified with the corpuscles of the cathode rays. We may, in fact, resolve the conception of electricity into that of matter, or the conception of matter into that of electricity. It is impossible to say that one of these conceptions is more fundamental than the other. Owing to our muscular sense, the forces required to set matter in motion are direct sense-perceptions, and thus to our minds matter seems more familiar than electricity ; but, had we an electrical sense, as the fish torpedo may possibly possess, we might find the positions reversed, and think it simpler to paint our picture of nature in electrical colours. Thus clearly we see how the structure of our minds, nay of our bodies too, affects our scheme of science.

Another consequence of the theories we are considering is of philosophic interest. A corpuscle or electron, as we have seen, owes its apparent mass to the field of electromagnetic momentum it carries with it. Now, while this momentum remains constant at all ordinary speeds, the mathematical equations show that it should increase very rapidly as the velocity of light is approached. If that velocity could be reached, the apparent mass would be infinite, for no exertion of force could cause any further increase in velocity.

Now, in the next section, we shall describe the phenomena of radio-activity in which corpuscles are found moving with speeds which in some cases approach to within some five per cent. of the velocity of light. By measuring their deflection by electric and magnetic forces, the mass of these particles can be measured, and it is of great interest to find that once more our scheme of science corresponds with observed phenomena ; the measured mass of the corpuscle shows the calculated increase.

Thus, at speeds such as these, the constancy of mass, one of the fundamental axioms of the Newtonian dynamics, fails us. A new dynamic must then come into action. We shall be brought back to this remarkable conclusion later by another road.

Both cathode rays and Röntgen rays produce luminosity when they strike phosphorescent screens. Radio-activity. This phenomenon suggested the idea that phosphorescent substances might themselves emit similar rays, and in 1896 led to the discovery by Henri Becquerel that compounds of the metal uranium, whether phosphorescent or not, constantly emitted rays which affected a photographic plate through opaque screens, and rendered gases through which they passed conductors of electricity. In 1898 M. and Mme. Curie, finding that the mineral pitchblende was more radio-active than its contents of uranium suggested, succeeded in isolating from it compounds of an intensely active new element, to which they gave the name of radium.

Radio-active substances emit several types of radiation. Two of these certainly consist of projected electrified particles, which are deflected from their straight paths by the action of electric and magnetic forces, to an extent which shows that, while one type is identical with the sub-atomic corpuscles of cathode rays, the other consists of a stream of particles of atomic dimensions, now known to be atoms of the gas helium.

The emission of rays is always accompanied by

chemical change, new substances appearing. Thus radium evolves continuously a radio-active gas or emanation, which itself deposits radio-active matter on solid surfaces. Moreover, these changes are associated with the emission of large amounts of energy, which may be measured as heat, and are shown to transcend by far the energy liberated by any ordinary chemical action.

These striking phenomena are explained and co-ordinated by a theory propounded by Rutherford and Soddy in 1903. Radio-activity is due to the explosive disintegration of individual atoms, one in many millions breaking up each hour. In this way it is calculated that half of a given quantity of radium would disappear in some two thousand years, passing successively into the emanation and various kinds of solid matter, while each act of disintegration involves the explosive emission of one atom of helium. Radium itself is probably formed by a similar even slower change from a substance now called ionium, which itself is derived directly or indirectly from uranium. Uranium thus becomes the first known ancestor of the family of radio-active elements of which radium is the most distinguished representative.

In this way the dream of the mediæval alchemist at length has come true. Chemical elements are not all indestructible and eternal, though their changes seem to be beyond our control. We can but watch one type of matter arising from another ; no philosophers' stone will produce the change ; even the resources of a modern laboratory seem powerless to hasten or delay the disintegration of a single atom.

While all this new knowledge has expanded amazingly the bounds of science as conceived thirty years ago, in many ways it has but served to confirm and give confidence in some of the fundamental concepts in which that science was expressed. In especial, the atomic theory has come triumphantly out of the ordeal, and has proved its continued value as an interpreter of nature at the present stage of the advancement of learning.

The Individual Atom.

Till recent years, the theory that matter consists of an enormous number of discrete molecules and atoms rested on physical and chemical evidence of an indirect character. The effects of the individual molecule or atom seemed for ever beyond the power of direct perception. We could but infer them from the effects of many millions acting together, as the scout in an airship at a high altitude might infer the presence of individual soldiers from the movement of the columns of an army on the march. Even after the striking success which attended the application of the atomic theory to the co-ordination of the phenomena of electric conduction in liquids and gases and of radio-activity, there were not wanting sceptics who held that individual atoms must remain purely conceptual entities, removed from the possibility of direct sense-perception.

But single atoms of helium, shot off by radium as a rays, have been revealed in two ways. Each atomic projectile produces a long train of electric ions as it passes through a gas before its energy is exhausted, perhaps by knocking loose corpuscles out of the molecules it encounters in its path. These ions have two effects. They make the gas a conductor of electricity, while they exist, so that, by placing the gas in circuit with a battery and an electrometer, Rutherford has shown the effect of each α particle by the sudden throw of the needle of the instrument. Secondly, the ions act as nuclei for the condensation of mist, and, in this way, C. T. R. Wilson has made visible as a line of cloud the track of each particle.

As we have seen, the revival of the old theory of an all-pervading æther was due to the researches of The Lumini- Young and Fresnel, who re-established ferous Æther. the wave theory of light. If light be waves, something to undulate must exist through space. Maxwell's electrical theory demanded a similar something to carry electromagnetic waves, and strengthened greatly the tendency to assume that this something was a subtle, material or quasi-material substance, extended permanently throughout all space, linking together the most distant stars, and capable of being described in mechanical terms.

It was, however, clear at once that æther must have properties unlike those of any other type of matter known to us. Light-waves oscillate in directions at right angles to the line of sight. Hence the medium which carries them must possess elasticity of shape. In this it must be analogous to a solid. Yet, since it offers no appreciable resistance to the motion of planets, it must be not only fluid but excessively attenuated. Reconciliation between these results was sought in different ways. For instance, a medium was invented rigid to very rapid movements, but offering no resistance to movements comparatively slow. Æthers followed one another fast—our

16

conceptual space became filled with turbulent motion, with interlinked vortex filaments, even with solid, densely packed spheres.

But worse difficulties remained. If space be filled with a stagnant æther, it must either be unaffected by the movement through it of ordinary matter, or the moving matter must disturb it, and carry some of it along in the path of motion. For instance: was the æther near the surface of the earth moving with it, or was it at absolute rest in space? A test was possible on the velocity of light. Was that velocity the same in the direction of the earth's motion as across it? In a famous series of experiments, Michelson and Morley found no relative movement between the earth and the æther, and concluded that the earth swept an atmosphere of æther along with it.

On the other hand, Lodge, who sought to detect the drag of the æther by whirling masses of steel, found no effect. However fast the steel moved, it left the æther just outside it at rest. Similar discrepancies came to light in other optical problems.

An explanation of the result of the Michelson-Morley experiment was offered simultaneously by Fitzgerald and by Lorentz. If matter be an electrical manifestation, the forces between its particles, and therefore its dimensions, might change as it moved through the æther, and this change of size in Michelson and Morley's apparatus might just compensate for the effect they sought.

But whether this explanation were right or no, the authority of the æther remained supreme. No one doubted the fundamental validity of the conception. Indeed, attempts were constantly made to express matter in æthereal terms. Lord Kelvin's idea of vortex rings in a perfect fluid, Karl Pearson's view that matter might be æther coming into and going out of three dimensions from and to an unknown space of four dimensions, Osborne Reynolds' mechanism of chasms in a medium made of hard spheres packed closely together, Larmor's picture of an electron as a strain knot in the æther—all are attempts to explain matter in terms of æther, and thus to bring unity into physical science.

In spite of the incompleteness of these attempts, few doubted that success was but a question of time. The splendid vision of a conceptual Universe, framed in its physical aspect out of the known properties of one all-pervading medium, led men on. It seemed that the vision might be realized any day, and all science become but the physics of the æther. Materialists who wished to be specially forward in the stream of progress began to call themselves æthereal-ists. Whether they were wise time alone can show.

The difficulties which surround the theory of a stagnant æther at absolute rest in space have been evaded
The Principle by a new principle evolved in different of Relativity. ways by Einstein and by Minkowski. By reason of the failure of all attempts to measure absolute translatory motion through space by means of reference to the æther, Einstein concludes that it is a principle of nature that such absolute motion can never be detected. The only translatory motion that has meaning for us is the relative motion of one body and another. To use this principle as a basis of deduction, Einstein also assumes that the

velocity of light in free space appears the same to all observers, regardless of the motion of the observer or of the source of light.

These two statements taken together form what is called the Principle of Relativity. They are in accordance with all facts known at present, and, although they lead to unexpected and startling results, none of those results have been yet shown to be inconsistent with each other or with the results of observation and experiment.

If we imagine two bodies to be moving past each other in free space on parallel paths, as two railway trains may pass each other on opposite lines, a simple geometrical construction will convince us that, whichever body we assume to be at rest and whichever in motion, on the principle of relativity, the unit of time on the moving body is to the unit on the stationary body as unity is to $\sqrt{1 - v^2/c^2}$, where v is the velocity of the body, and c the constant velocity of light. Thus, to an observer on either body, the time scale of the other body seems different from his own. Similarly, there is a corresponding change in the unit of length, a change equal to that suggested by Lorentz and Fitzgerald.

When we pass to dynamics, we find once more that the principle of relativity requires that the mass of a moving body should vary, and, in terms of the mass of a stationary body, should be $1/\sqrt{1 - v^2/c^2}$. If the body moved with the velocity of light, v would be equal to c, and the mass would become infinite. Here again, the principle leads to the same result as the ordinary electromagnetic theory, and is in equal accord with the results of experiment.

It will be seen that, on the principle of relativity, all attempts to measure the velocity of anything relatively to a stagnant æther are foredoomed to failure. Compensating adjustments in the standards of time, length and mass conspire together to defeat us.

In fact, the attempt to investigate the æther itself may have to be abandoned. It may be that in inventing " a nominative case to the verb to undulate " we have gone too far in imagining an all-pervading stagnant medium with the mechanical or quasi-mechanical properties suggested to our minds by the too vivid words " universal æther."

We can but feel that once more the self-sufficient structure of nineteenth-century science has sustained a rude shock. It may recover therefrom, and the universal stagnant æther once more surround us. Or that section may fall and have to be rebuilt on newer lines, and the universal stagnant æther pass into the great historical museum of intellectual curiosities. It will be in good company, and may wait in patience, with the firm hope of welcoming in due course the conceptions which now may supplant it, and may even cherish the possibility of a triumphant reappearance on the acting stage of science.

Simultaneously with the later work of Darwin (1865), a series of researches was being carried on in the cloister of Brünn which, had they come to his notice, might have modified the history of Darwin's hypothesis. Gregor Johann Mendel, a native of Austrian Silesia, an Augustinian monk, and eventually Abbot or Prälat of the Königskloster, not satisfied that Darwin's view of

Mendel and the Laws of Inheritance.

natural selection was sufficient alone to explain the formation of new species, undertook a series of experiments on the hybridization or cross-breeding of peas. He published his results in the volumes of the local scientific society, where they lay buried for forty years. Their rediscovery, confirmation and extension by William Bateson and other workers marks the first step in the recent development of heredity as an exact experimental and industrial science.

The essence of Mendel's discovery consists in the disclosure that in heredity certain characters may be treated as indivisible and apparently unalterable units, thus introducing what may perhaps be termed an atomic conception into the field of biology. An organism either has or has not one of these units ; its presence or absence are a sharply contrasted pair of qualities. Thus the tall and dwarf varieties of the common eating pea, when self-fertilized, each breed true to type. When crossed with each other, all the hybrids are tall and outwardly resemble the tall parent. Tallness is therefore said to be " dominant " over dwarfness, which is called " recessive." But when these tall hybrids are allowed to fertilize themselves in the usual way, they are found to be different in genetic properties from the parents whom they resemble outwardly. Instead of breeding " true," their offspring differ among themselves ; three-quarters are tall and one-quarter are dwarf. The dwarfs in turn all breed " true," but of the talls only one-third breed true, giving rise solely to tall plants, while in the next generation the remaining two-thirds repeat the phenomena of the first hybrid generations, again giving birth to pure dwarfs and mixed " talls."

These relations are explained simply if we suppose that the germ cells of the original plants bear " tallness " or " dwarfness " as one pair of contrasted characters. When a tall plant is crossed with a dwarf one, all the hybrids, though externally similar to the dominant parent, have germ cells half of which bear " tallness " and the other half " dwarfness " in their potential characters. Each germ cell bears one or other quality, but not both. Thus when, by the chance conjunction of a male with a female cell from these hybrids, a new individual is formed, it is an even chance whether, as regards the qualities of tallness and dwarfness, we get two like or two unlike cells to meet ; and, if the cells be like, it is again an even chance whether they prove both " tall " or both " dwarf." Hence, in the next generation, we get one quarter pure " talls," one quarter pure " dwarfs," while the remaining half are hybrids, which, since tallness is dominant, resemble the pure " talls," and in appearance give three-quarters of the seedlings that character.

It will be seen that the methods of inheritance are different in the cases of dominant and recessive characters. While an individual can only transmit a dominant character to his descendants if he himself shows it, a recessive character may appear at any time in a pedigree if two individuals mate who carry the recessive character concealed in their germ cells, though not outwardly visible in themselves. But in the majority of cases the conditions of inheritance are far more complicated than would appear from the study of two simply contrasted qualities in the green eating pea. For instance, qualities may act as dominants or recessives according to sex or other conditions. Characters may be linked in pairs so that one cannot appear without the other, or again they may be incompatible and never be present together.

Many Mendelian characters have now been traced in plants and animals ; while, as a practical guide in breeding, the method has been successfully applied to unite certain desirable qualities and to exclude others of harmful tendency. By this means, Biffen has established new and valuable species of wheat, in which immunity to rust, high cropping power and certain baking qualities have been brought together in one and the same species owing to a long series of experiments based on the Mendelian laws of inheritance.

The extension of this new method of research to mankind at once cleared away many old puzzles and opened up fresh fields of study. Many diseases and malformations have been proved to be dominant Mendelian characters ; while deaf-mutism, and other defects, appear to be recessive, and therefore especially liable to appear in cousin marriages between sound individuals who come from families liable to the affliction.

Of normal healthy characters, eye colour alone has yet definitely been proved to show Mendelian phenomena in its line of descent, brown colour being dominant to gray, though indications are not wanting that further analysis will disclose similar relations in many other directions.

The application of such results is obvious. We can already predict accurately the probable results of human matings when the potential parents possess a character which has proved to be Mendelian in its line of descent. Increase of this knowledge is coming fast, and should soon form a valuable guide to the physician, to the sociologist, perhaps to the statesman.

From the more philosophic aspect, the results of Mendel and his rediscoverers complete on one side the Darwinian period, and on another show its insufficiency, and open a new chapter. In the original theory of the origin of species by natural selection, still held as against the Mendelians by many biologists, it is the small variations of organisms round the mean form that supply the material for selection to work upon. It is not easy to see how enough isolation of more successful beings can occur to establish inbreeding, and so build up a new species. But the large differences found by Mendelians, and the complete separation of unit characters, do much to diminish this difficulty. In their view, new varieties may arise almost *per saltum*, and establish themselves at once.

On the other hand, these new and unexpected phenomena showed that Darwin, as he himself well knew, had not said the last word on evolution, and gave a welcome check to some of the more confident speculations of the newer prophets of Darwinism.

Another aspect of evolutionary philosophy also has been affected by recent research. Though Lamarck's theory gave up its pride of chief place to that of Darwin, many accredited evolutionists continued to preach and act as though characters in man acquired by the reaction of the environment—as by exercise, training, or education—were handed down in their developed form to his offspring. Thus a comforting doctrine arose that we need but improve the acquirements of one transient generation for the next to be better innately—that the race by mere well-directed philanthropic enterprise would continue to improve indefinitely in mental and moral worth. Then came Weismann, who, led by the evidence that the germ cells of one individual are derived by direct descent from those of his parents, asked how characters acquired during life could affect the germ cells present almost before life began? He examined critically every suggested case of the inheritance of such a superimposed character, and found each case to crumble away under his analysis. We have not reached certainty. Some indications are believed to point to possible though rare occurrence of partial inheritance in a few insects and plants when modified by the environment. But, as a chief actor in the hereditary drama, the acquired character is discredited.

Here again we touch the problems of sociology. Though political and social institutions acquired by one generation are certainly inherited by the next, and if well suited to the natural development of the people may help them progressively to advance in social organization by a process of cumulation, this inheritance of social organization is not evolution in the Darwinian sense. And, in the far more important and fundamental inborn qualities of the race, no rise of one generation by improved hygiene, exercise or education can affect, save indirectly, the qualities of the next. Selective parenthood, natural or conscious, is alone capable of raising our race or preventing its degeneration. In the light of the indications of danger referred to on p. 229, these considerations are of momentous import.

But it is time to turn to other biological tendencies, in which, perhaps, even greater philosophic interest lies. Here too we have to trace an old story in a new form. As the premature determinist mechanical theory of life of the Greek atomists gave way before the equally premature vitalistic dogmas of Aristotle, so the still premature mechanicism of the too ardent physiologists and Darwinians of the nineteenth century shows signs of yielding for the time before the continued analysis of the problem of life. Charles Darwin himself, be it noted, kept an open mind on questions authoritatively answered by ultra-Darwinians, especially in Germany.

Neo-vitalism.

The success of physiologists in co-ordinating many functions of living organs with physical and chemical changes led to a belief that life itself could be stated completely in physical and chemical terms, and that organic beings were but more complicated inorganic machines. The fact that all living changes cannot yet be explained by known physical principles is a poor argument against a mechanical philosophy, for it is supremely foolish to found our faith, in any department of human thought, on mere gaps in knowledge, some of which, at all events, can only be temporary.

Each year sees an increase in the number of bodily functions which can be described in physical and chemical terms. In particular, the chemical action of certain internal secretions poured into the blood is being found to influence the functions of distant cells, and thus to control such characters as stature, the movement of the heart, even the development of the brain.

But, nevertheless, some biologists have held that

recent work leads away from and not towards a complete and satisfying explanation of many vital processes by physical principles. The organism, it is argued, uses physical processes, as indeed is obvious, but it does not necessarily, or even probably, follow that the controlling agent itself is purely physical. Moreover, in comparing a living being with a mechanical machine, it must be remembered that the machine " is no ordinary sample of the inorganic world.   It is an elaborated tool, an extended hand, and has inside of it a human thought.   It is because of these qualities that highly complex machines come to be so like organisms.   But no machine profits by experience, nor trades with time as organisms do."

Indeed, many biologists, such as Driesch, Wolff, Bunge, and J. A. Thomson, hold that the essence of the problem is overlooked when attention is focussed solely on the details of organic life, and the mechanical, physical and chemical processes used therein.   From a more complete survey, they see reason to believe that there are facts concerning organisms considered as wholes which are not covered by any possible extension of the domain of mechanics, physics and chemistry.   A living being, they believe, is autonomous ; it is in a sense an ultimate unit in nature, and cannot be explained by a study solely of the detailed phenomena of its parts and organs.   Besides the obvious purposefulness of the higher animals and plants, a similar intention appears throughout all living beings even from the first.

As an instance of the phenomena on which they rely, we may describe one of the many results of recent work on the process of egg-development—" Entwickelungsmechanik," as it is not very happily called. Driesch found that if the essential part of the egg of the sea-urchin (*Echinus microtuberculatus*) were separated into two cells, each cell divided and developed in the normal way. For some stages, the product of development resembled what would have been obtained by cutting in half the product of a complete egg, but soon the hemisphere became rounded into a sphere, and a whole larva, smaller than usual, but complete in all its parts, ultimately developed.

Such phenomena as these, the facts of adaptation to changes of environment, the restitution of injured parts in adult organisms, acquired immunity, the reproduction of complex beings by the hereditary process, above all the marvellous properties of mind and consciousness studied by psychology, cannot be put on one side. They make it impossible for a growing number of biologists to regard life as a cumulation of chemical and physical changes, and have gone far to re-establish biology as an independent science. They have led to the revival of the old hypothesis of vitalism, or, as some now prefer to call it, of the autonomy of life—an old theory with a new name, arising refreshed from its years of retirement with stronger claims to notice.

Still, the majority of biologists remain convinced that physiology in its essence is but applied physics and chemistry, and that, although we are yet far from a complete account of the organism, it is ultimately explicable with no reference to unknown principles, such as vitalism, old or new. Whatever be the truth, it is probable that this attitude will inspire much of the work of the coming years. It must not be forgotten that, even if the whole of life regarded externally as a subject of scientific study be reduced to physical or mechanical terms, the internal phenomena of consciousness, when the mind looks inward on itself, remain a world apart.

The extreme Darwinian position — a position Darwin himself never claimed—has been criticised by Wigand, Nägeli, Wolff and others. Natural selection is allowed by all to be a *vera causa* in determining the limits of species. But it is now pointed out that it can only eliminate what cannot survive, never create diversities. To explain by natural selection the existence of some organic character is, says Nägeli, as though one explained the presence of certain leaves on a tree by saying that the gardener had not cut them away. That statement may explain why other possible leaves are not there, and why only the actual leaves to be seen have survived, but behind it is the marvel of why—gardener or no gardener— there are leaves at all.

Natural selection, it is said, fails to account for mutual adaptations, such as those existing between plants and insects. It fails too to explain organs which are composed of many parts, such as the eye, and still act as functional units. It fails once more to account for the first origin of organs, which only become of survival value at a later stage of development. Through all life appears an inherent purposefulness, as an essential and fundamental property of the organism. This evidence of purpose, of direction, of intention, underlies the whole of biological development, and may give the clue to a revaluation and extension of the facts now available for discussion.

While the relationships and similarities between different species point to a common origin, and enable us to assert a near connection between certain types, such as man and ape, the facts seem to warrant little else. We cannot tell what really happened in former ages, when, perhaps, the stream of protoplasm was more plastic and more variable than it is to-day. The elaborate theories of descent which have successively traced the origin of the vertebrates to amphioxus, worms, spiders and crayfishes, have been compared in scientific value by Emil du Bois-Reymond with the pedigrees of the heroes of Homer.

If we accept the conception of descent, the theory of evolution—and no one proposes to jettison it—the evidence of purposefulness in living beings suggests that some unknown principle of organization must have been at work from the very beginning. Here again we are brought back to neo-vitalism.

Whether these vitalistic tendencies will persist, and crystallize into a consensus of opinion, it is impossible to say. The analogy of the past points to the probability that here too we are but watching a passing phase in the rhythmic wave of human thought. Perhaps some new " explanation " of vital phenomena may be offered in the years to come, and once more give rise to a complete mechanical conception of life. Once more mechanism may fail when subjected to a closer scrutiny, and a still newer vitalism come to its own. Let us at least hope that each alternation may mark a step in knowledge, and lead to the more exact formulation of fundamental problems which as yet we can but dimly discern. The historian of thought tends to be sceptical about ultimate solutions. Yet to the scientific protagonist at each stage, an animating faith in the eternal verity of his own creed, whether of vitalism or mechanicism, is a priceless possession. A man without convictions cannot be convinced. Let him therefore face his facts fairly and choose his line. As long as he does not try to fit his conclusions to any preconceived ideas of what may be good for man to believe, he will play a useful part ; but, " die Philosophie muss sich hüten erbaulich sein zu wollen."

Introspective psychology is at least as old as the Greeks. Experimental psychology, and comparative Modern psychology, the study of the minds of Psychology. different races of men and animals, are among the newest born of the sciences.

Experimental determinations can now be made of the sharpness of the senses, and the action on them of injury, fatigue and other changes ; of the power of memory and the effect of association of ideas on memory and on action.

The human mind possesses the special faculty of constructing complex ideas and following trains of thought. Here too association comes into play ; association not only of sense impressions which have followed each other in the past, but also of things similar in various ways but perhaps never before associated. In this formative process of comparison the importance of language becomes manifest. The possibility of giving all similar things a class-name helps the mind to frame a universal or general concept of that class. To call all dogs dogs, and not merely Joe, Nelson, Cæsar and Smut, is a very real mental achievement. To the savage, indeed, this power of classification is so mysterious that from the spoken word, to us indicating a class association, springs the idea of a mystic or divine unity, binding together in blood-relationship all the members of that class. Hence comes *tabu*, and perhaps a tutelary deity. The continued fascination of the concept of class is seen passing from Plato and Aristotle to the Scholastics of the Middle Ages.

Besides our conscious mental processes, a subliminal mind may be traced, working " sub limine "—beneath the threshold. At length the phenomena of trance, of hypnotism, of suggestion, of multiple personality, are being submitted to scientific examination. The results of these researches are not confined to the special states inquired into ; they throw light also on our normal psychological processes and are intimately connected with the phenomena now being investigated in " psychical research." To some observers recent experiments appear to reveal direct thought-transference from mind to mind, and even to suggest the continued existence of the spirits of the dead, and the possibility of difficult, but still intelligible, intercourse between them and the living. We seem to return to the " aerial " and " terrestrial " states of the soul, outlined by the Cambridge Christian Platonists of the late seventeenth century, and to be testing some of their theories of possible intercommunication between the two conditions.

The realization of the importance of conscious and subconscious association of ideas is helping us to understand in a new way many dark phases of human life and society. We find that our social, political and religious actions are profoundly affected by association. In the eighteenth century a general intellectualistic theory of politics and religion was developed by social philosophers, and in the nineteenth it was applied freely in practice. Man was (so they assured us) or should be—save as to human imperfection—governed exclusively by the cool light of reason. Passion and prejudice were transient storms that merely ruffled the surface of the normally unclouded mind. A crowd was but a number of individuals, each a potential philosopher with the torch of Reason in his hand. The " People," could their real will be ascertained, would govern not only themselves but each other with the calm wisdom of the sage, the intellectual acumen of the statesman, and the irresistible claims to righteousness of a numerical majority.

Now this *a priori* intellectualism has been shattered by the patient study of facts. Men are not and never have been governed by reason. The number of things about which we, even the best of us, consciously reason is very small. Economy of thought is as important to success as economy of wealth. For the most part, education and training are but practice in association, either of ideas with ideas or of ideas with action. Indeed, as Dr Whitehead tells us, " civilization advances by extending the number of important operations which we can perform without thinking about them. Operations of thought are like cavalry charges in a battle—they are strictly limited in number, they require fresh horses, and must only be made at decisive moments." The simple association of twice two with four, the complex series of associations which enables a mathematician to solve a differential equation, are the same in kind as—though founded on a more general consensus of competent opinion, and therefore on a more legitimate basis than—the impressed associations of our youth, which taught us to identify Lord Beaconsfield and Peace with Honour, or Mr Gladstone's special variety of Peace with Retrenchment and Reform. Both are methods of saving thought, the one in solving equations, the other in deciding how to vote.

The modern arts of advertisement and of electioneering, which is but a species of advertisement, are the best examples of the application of psychological principles in practice. In each case, simplicity of association is of primary importance. The leading English newspaper, to which the arts of advertisement have become by no means unfamiliar, in a recent anniversary number, lays bare the psychology of the method :—" The iterated appeal to the eye stamps its form on the brain without any special connotation, and in the absence of any opposition remains there until some favourable occasion or recollection brings it into practical association with the wants of some individual." If we are continuously impressed with the simple connection between Pulley's Moonshine Soap and Silver Purity, some day we forget to make the mental effort needed to realise that all the bills our eyes have seen bear the impress of one man's desire—and order Moonshine. If the political candidate, who appeals most to the innate bent of our mind as modified by environment, or to the hereditary political opinions bequeathed by our forefathers, descends on our constituency but once in a while to dispense lavish hospitality with the

simple message of, " Tax the land and vote for Jones,"
or " Tax the foreigner and support Smith," we shall
find ourselves much more enthusiastic in his support
than if he lives among us and confuses his electioneering virtues with his qualities, perhaps his faults,
as a neighbour.

Then the psychology of a crowd has points of difference from those of a number of isolated individuals—the
management of a public meeting is not the same art
as that of the skilful canvasser. The speeches of
Brutus—the doctrinaire man of reason—and Mark
Antony—the astute and emotional opportunist—in
Shakespeare's *Julius Cæsar* are marvellous examples
of an inborn knowledge of crowd psychology. Every
would-be platform speaker should know them by
heart.

Now the nineteenth-century theory of democracy
was founded on the special varieties of intellectualism
then prevalent, reinforced by Lamarck's view of the
inheritance of acquired characters. Unfortunately,
we know now that, owing to the essential psychological
structure of the human mind, public opinion can be
manufactured, and is manufactured daily, without
any regard for reason. The long purse, the unscru-
pulous or misguided conscience, the oratorical appeal
to passion, prejudice and self-interest, have greater
power of creating public opinion than careful reasoning based on verified facts. Wherefore we cease to
wonder that in those countries which most pride
themselves on their democratic character the millen-
nium is no nearer than in others. The power of mere
benevolent stupidity in making history is a new
discovery. Stupidity is not necessarily due to evil

intent or false reasoning. It is as often caused by false associations, for which the intellect of the sufferer is not primarily or even chiefly responsible.

In constitutional problems, too, we are passing away from the purely intellectualist standpoint ; and once more the change began before its psychological import was understood. If we compare the attitude of mind of 1860 with that of to-day, we mark the difference. At that date the tendency was to a complete utilitarianism. A business-like constitution whose every part had its direct economic use was the ideal. The Monarchy was an interesting survival of a barbarous past, which few wished forcibly to remove, but for which few ventured to foretell a prolonged existence. The colonies were somewhat useless and expensive encumbrances, which might almost be encouraged to " cut the painter." Colour and glow were passing out of public life, and grey, useful uniformity was spreading over all.

Slowly a change began. It can be traced back through the " Oxford Movement " to the romantic and historic inspiration of the novels of Sir Walter Scott, an Anglian of the North British stock. Once more the idea of Nationality was found to appeal to sentiment, and to have driving force behind it. It was discovered that " the golden link of the Crown " stood for a corporate ideal of the State, which gained real strength thereby. It was less easy to glorify as the impersonation of the nation a bourgeois President in a frock-coat—still less was it easy to die at his behest should the need arise. Colonies ceased to be regarded as unprofitable branch establishments, and became the Dominions Over Seas where lay the homes of our kith and kin. Coronations regained their old splendour, and Jubilees were invented to supplement them. Even towns begin to realize a sense of corporate continuity, and pageants minister to the newly recovered or newly acknowledged processional instinct. We revive maypole dancing and introduce acting and dramatic recitations into the elementary schools—thereby returning to the educational ideals of the later Stuarts—and think we have taken a step in advance.

Now all this movement has a real historical importance in the development of the human mind. It accompanied the contemporary change of attitude in science, and was its representative in another field of thought. The sages of 1860 had overlooked essential psychological truths, and had made man a purely intellectual being in their own fancied likeness. Since then we have rediscovered his other faculties, and have set to work, some of us to develop them, others to find an equation which will express their value in the corporate life of the community.

Meanwhile, the application of comparative, historical and evolutionary methods to the study of religion has changed profoundly our ideas of its course of development.

Comparative Religion.

Even when the idea of evolution had been absorbed and assimilated by theologians, the *a priori* method reigned for a time supreme. Evolution had proceeded, no doubt, but dispensation had followed dispensation, till, in the fullness of time, the particular stage of Catholic doctrine now authorized by the Holy Father, or the particular tenets then emphasized by the

Protestant sect favoured by the enquirer, had emerged and proved their adaptation to the environment. Heathen lands might still be painted black on the missionary map, though it began to be realized that there were shades of grey between black and white, which required explanation and apology.

One underlying assumption made by theologian and sceptic alike vitiated the whole conception of the evolution of religion—an assumption inevitable till the lesson of patient study of primitive peoples had been learned. It was assumed that, at all stages, creed and dogma were the essence of religion, and rite and ritual but their expression.

Now the recent researches of anthropologists fail to accord with this preconception—as, indeed, so often facts do fail to accord with *a priori* theories.

Far earlier than definite theological beliefs or dogmas are religious rites and ritual observances. It is a comparative advance to " bow down to wood and stone." When the really primitive man wants rain or victory he does not ask a god to give it to him, but goes and tries to secure it himself. When it rains, the thunder sounds or the frogs croak. If that be all, he can do it too. So he whirls his bull-roarer to make thunder, or hops and croaks like a frog to bring the rain. So rises magic, " that spiritual protoplasm," as Miss Jane Harrison calls it, " from which religion and science ultimately differentiated." Then come ritual, magic dance, and outpourings to fertilize the earth.

Social observances, necessary for the complicated savage life, become inextricably mingled with these nature-spells. A boy must leave the women who tended his childhood and join the men in hunting and fighting. The break is essential, and must be effectual for the social well-being of the tribe. Hence spring initiatory ceremonies, often of cruel severity, and revelations of tribal secrets and *tabus*. Here too is a fertile source of ritual, acquiring all the sanctity of custom. Leaders in ritual ceremonies tend to become medicine men or priests. In time perhaps they acquire divine attributes.

At this point another parallel development of ideas is ready to join hands. To the savage, memory, anticipation, dreams, give worlds as real as that which alone is held by civilized man to be objective. All things grouped under a common name become related in blood—need a common origin. Tutelary deities, divine ancestors, priests, spirits of wood, mountain and stream emerge into vague consciousness. In terms of these conceptions the familiar rites and ritual tend to become explained. In a bewildering variety of imagery religion rises for mankind.

It will be seen how far these facts of primitive life are from the intellectualist theories of religious evolution current in the nineteenth century. In the light of those theories, man believes, and therefore there are rites and sacrifices. But primitive man, at all events, has rites and sacrifices, and therefore believes. Then, it is true, rite and sacrifice acquire an added sanction, the process becomes cumulative.

Like politics, religion is seen to be affected profoundly by the psychology of the race. The development of the religious faculty in man is carried on by his whole being, mental and emotional, not by his intellect alone.

It is probable that, when the significance of the new

knowledge of comparative religion is grasped by the world at large, men, undeterred and unwarned by experience, will feel their faith in danger and the fountains of the deep unloosed upon them. When they find how many cherished doctrines and how much beloved ritual arose from magic and nature-worship and were common possessions of many faiths ; when they see that on those rites and doctrines Christianity was merely grafted to supply a new interpretation of the mysteries, they will feel religion itself is crumbling before their eyes.

Yet, in truth, it is but the old story of the sun's place in nature and the old story of evolution. When natural selection gave a comprehensible theory of the method of creation, some men rushed to the conclusion that life and all existence were but the by-play of materialistic mechanism. Slowly they came to see that things stood much as they were, save that a fresh revelation of How had been given to mankind.

So, in this new field of knowledge, a revelation of the method and process of the development of the religious experience of mankind does not alter the fact of its existence, or make shallow the depths of the soul's sea of awe and reverence for its own life and its intuitive apprehension of the divine. The Kingdom of Heaven is still within.

The words of Shakespeare are to be found in a dictionary common to all mean books. That does not destroy the genius of the poet. Though the ritual and dogmas of Christianity, as of other religions, lie scattered in a thousand creeds, the conception of the Divine mind that reinterpreted and reinterprets them to all ages is not less Divine. When we understand the origin of our Church, the derivation of its sacraments, the history of its growth, the majesty of the living structure is no whit diminished. Once more we have learned something about methods. Though the Church may have evolved by natural means, that should teach us to see God in Nature, not to deny the Church to the God of Nature. Though religion arose naturally and inevitably by the psychology of man, that should show us that man's whole nature is essentially religious.

Attempts to support religion on an intellectual assent to dogma alone are doomed to failure. They ignore the essential psychological basis of the religious instinct. Churches and ritual are a real need to most men, as the Oxford Movement rediscovered. To one the gorgeous but direct and simple imagery of Rome is necessary ; another is best stirred by the more refined symbolism of a plain white surplice and the majestic diction of the Book of Common Prayer ; while to a third the profound psychological device of the crowd-silence of a Quakers' meeting is the most effective and impressive as it is the most subtle, and psychologically the least simple, form of ritual. An effective Church must be able to enshrine the glamour of the past in continuity of rite and tradition, to meet, by continual routine of prayer and praise, man's psychological need of association in the development of the religious sense ; and to satisfy the artistic and emotional sides of man's nature in dignified ritual, while maintaining its hold on the best spirits of the age by an open-minded receptivity to intellectual progress. A nation that forsakes its Church, or suffers its Church to forget traditions on the one side or to fall into unreasoning obscurantism on the other, a nation that ignores the survival value of religion, will perish from off the face of the earth.

Alongside the development of physical and biological science which we have traced in the preceding pages, new attention has been devoted to the fundamental concepts which lie at the base of all natural knowledge. Ernst Mach in Germany and Karl Pearson in England have modified and developed the doctrines of Mill and Herbert Spencer.

Theory of Scientific Knowledge.

In the laboratory, as in the practical life of the field or the office, there is no time for philosophic doubt. A naïve realism is alone possible. We assume that we know all about our spade or our pen, our test-tube or our atom. Can we not see them, handle them, or, at all events, trace their effects ? Yet if we think carefully, we shall see that the case is not so simple as, for practical convenience, unconsciously we assume.

A spade is to our eyes a long-shaped figure of a brown colour, tipped with a flat blade of grey. The stored impressions of memory enable us to say it is of wood and iron, and thereby to endow it with the qualities we associate with those materials. We construct a mental image of the spade, partly by sense-impression, partly by an unconscious act of memory. Our image is real, we can reason about it, develop it by trying experiments and adding to our ideas of wood and iron an idea of the complete spade as an implement useful for digging, or, at a push, as a weapon of offence. We are not conscious of our mental process of synthesis ; we regard the spade as a real " thing-in-itself " outside us, resembling in its ultimate character our mental picture of it.

But what right have we to assume the real existence of such an outside object ?   What is it that we really know when we observe the spade ?   Nothing, in effect, but impressions on our senses.   Our optic nerves send certain stimuli to the brain ; we call them sight of a long brown object with a grey blade.   We stretch out our hand and seize the spade.   Yet all that we expect, and all that we experience, are certain tactual sensations given to the brain by our sense of touch.

Chemists have resolved the spade into atoms of carbon, hydrogen, iron, etc.   Physicists have gone further, and pictured the atom as a system of whirling electrons.   All these attempts to get at a more ulti- mate reality, merely mean that the sensations we experience at first may be changed into others by appropriate action, and that, to represent those new sense-perceptions to our minds, new concepts are needed.   By putting the spade into the fire, or by passing an electric discharge through the gases which are thereupon evolved, new phenomena appear, and a new mental scheme is evolved.   We cannot anyhow get at the external " thing-in-itself " ; we can but modify our sense-impressions and our concepts in certain limited ways.

Our brain has been likened to a telephone exchange in which the operator sits for ever locked.   His only knowledge of the external world is derived from the messages he gets or intercepts as they pass over the wires.   He may infer the existence of outside objects, he cannot prove it, and he can only gain a picture of those objects as revealed by the limited number of wires which enter the office, and by the kind of message which can come over the wires. He can hear, but he cannot see or feel. It is safe to predict that his mental picture of the world would be incomplete, and different from that which would be disclosed to him could be unlock his door and pass outside into the glory of the sunshine. How many more senses may be possible than our poor five, and what picture of our world would they reveal? But still it would be a phenomenal world, we could but obtain the record of our senses, never could we apprehend directly the " thing-in-inself."

Similar caution is needed in dealing with what are called laws of nature or natural laws. The word law in the legal sense has a meaning different from that which it ought to bear in science, where it should mean simply a shortened or convenient way of describing general routines in sense-impressions, or general relations between different mental concepts. When we say that the sun's visible rays are accompanied by warmth, we mean that we always find one sense-impression followed regularly by the other. When we say that two similar electric charges repel each other in the inverse ratio of the square of the distance, we assert a certain relation between our mental concepts of electricity, of force and of distance—concepts which have been developed by our minds out of the material of sense-impressions. The relation is suggested by experiment, that is, by sense-perception, but, once apprehended, it is used to develop a mathematical theory of electric forces and their effects in a purely mental or conceptual sphere. How far this complex scheme agrees with facts, *i.e.* with new sense-impressions, is a matter of experiment again.

The universality of natural law may, says Pearson, really be relative to the human minds involved, which, like machines that work for no coin but a penny, may sort out and analyse, all in practically the same manner, the material they will alone accept. This return to the doctrine of Kant—" the Ego prescribes its own laws to nature "—would avoid the assumption that these routines of sense-impressions, these relations between mental concepts, are produced by the unknowable, whether the unknowable take the form of matter, the thing-in-itself of the materialist, or the form of an immediate action of the Deity, the idealist reality of Bishop Berkeley.

Thus, to these empiricists, " science—refusing to infer wildly where it cannot know, and unwilling to assume new causes where the old have not yet been shown insufficient—treats the ' dead matter ' of the materialist," the mind-stuff of the idealist metaphysician, the immanent Deity of the natural theologian, " as a world of sense-impressions ; . . . the scientist . . . recognizes that the so-called law of nature is but a simple *résumé*, a brief description of a wide range of his own perceptions, and that the harmony between his perceptive and reasoning faculties is not incapable of being traced to its origin. . . . Our groups of perceptions form for us reality, and the results of our reasoning on these perceptions and the conceptions deduced from them form our only genuine knowledge."

On these lines the different sciences into which, for convenience, our studies are divided are but different sections cut through the model of nature our minds construct, or different aspects from which it may be regarded, or, to vary the simile, different-coloured flashlights thrown by our minds on the picture they have to examine. Mechanic looks at the model from one point of view, chemistry from a second, biology from a third. None is necessarily more fundamental in its essence than another, though mechanic may give a wider view, and from historical and psychological reasons seem to us more primary. Thus science as such is not only not committed to materialism—a belief in the dead matter of Moleschott and Büchner—as the sole reality, but does not involve a mechanical philosophy, as is so often supposed. Even should a theory of life come to be expressed in mechanical terms, and agree with observed phenomena, it would only show that the human mind found it more convenient to express its perceptions and conceptions of life in mechanical language. It would not show that there was any objective reality corresponding to the conceptual scheme, still less that that reality corresponded to the particular aspect of it which our minds selected.

On the other hand, this empirical scheme of science is equally far from lending support to an idealistic or theistic belief. It simply has nothing to say on the question of external reality, on the nature of " things-in-themselves." It purchases its freedom from both contending factions by a complete agnosticism, by a strict attention to its own business as conceived by itself—the attitude of the Liverpool Chinaman who, as a measure of security in a time of riot between Romanists and Orangemen, decked his shop-front with the legend, " Me no religion at all, me only wash clothes."

What then becomes of the eternal, immutable laws of nature, of which we have heard so often ? Are such laws non-existent ? By no means. But they are the logical laws of the conceptual world formed by our minds. Mathematics is in its essence symbolic logic, and mathematical and logical laws are of the same nature, though that nature is still a matter of dispute among philosophers. Some would hold here too an empiricist view, and say we gain a knowledge of such laws by experience. But, on the other hand, a rational school hold that logical principles are grasped by an intuitive action of the mind, and are not proved, though usually suggested, by experience. Once we understand the terms involved, we see instinctively that two and two are four in all conditions and at all times. Once we make the assumption that every particle of matter attracts every other particle with a force proportional to the product of their masses and inversely proportional to the square of the distance, the whole planetary theory follows logically, though it may need a Newton first to work it out.

Such relations follow from the structure of our minds, they are laws of thought. Some hold that they give us knowledge about a real world also—the world of universals, and are laws of nature as well as laws of thought. Such philosophers thus revive Plato's doctrine of ideas in modern guise. But however this may be, the laws enable us to build up a logical and necessary structure in the conceptual world once we have formed the conceptions and agreed on the definitions. How far this ideal world agrees in substance and in fact with the real world of sense-perception is an affair of experiment. Two and two always are four, whether the two and two be apples, or poets, or cats. Here the definitions are made to agree accurately with the sense-perceptions. But how far astronomical sense-perceptions will agree with the deductions of the Newtonian scheme is purely a matter of observation in the present and probability in the past and future. The fact that concordance exists over the period of two centuries during which our observations extend gives us confidence in the probability that we may extend our predictions backward into the past and calculate the dates of ancient eclipses, and forward into the future. But the farther we go, the less the probability of concordance becomes.

To be of practical use, all our conceptual schemes must thus enable us to predict the future behaviour of our perceptions. What ground have we for confidence that they will do so ? It is simply, as we have said, an affair of probability. There is no certainty in natural science. Because we have known the sun rise for ten thousand yesterdays, we frame a conceptual sun which rises regularly each morning indefinitely, and, as a matter of practice, we may bet ten thousand to one that the perceptual sun also will rise to-morrow. Because Newton's theory of gravity has met every demand on it for two hundred years, we have great confidence that the complicated conceptual astronomy founded on it by mathematical logic will still continue to agree with our observations of nature—that is, with our sense-perceptions. Because all chemical elements proved immutable for a hundred years, we assumed

18

that no chemical element could ever change—and in the surprising case of radium we proved to be wrong.

Let us return to the consideration of the empirical or phenomenal view of science—the view that it deals only with sense-perceptions and the concepts framed from them, and has no message about aught beyond. How far is that view right ? or, since we are here and now only historians, how far is that view accepted in the present, and likely to persist in the future ?

Probably most men of science who have thought about the subject would agree that science itself was thus limited ; that it could give of itself no certain information about the nature of any external reality. Its duty is to construct a consistent conceptual model, and to examine by observation and experiment how far that model conforms to sense-perceptions. But science, though it should be kept clear of metaphysics, as indeed our history abundantly shows, has much metaphysical import. When we leave science and take a metaphysical view, our science becomes one of the most valuable, perhaps the most valuable, of our sources of evidence. The empiricists are probably right in restricting science to sense-perceptions and mental concepts, but they are, perhaps, inconsistent with their own doctrine in inferring from the evidence of science that no other knowledge is possible, and metaphysics an empty dream.

In the earlier chapters of this book, we have found science treated as a branch of philosophy, with no Science and independent existence. Gradually we Philosophy. have traced its annexation of province after province from the realm of philosophy, its libera-

tion from philosophic bonds, and the establishment
of its right to sovereign jurisdiction within a kingdom
acknowledged as its own. Now, when we return to
the subject, we find, in a sense, the positions reversed.
Science stands on its own ground, alone, secure ;
philosophy has to frame its systems in conformity
with natural knowledge.

What, then, is the bearing of our present scientific
position, and of the tendencies we have traced in
recent scientific theories, on philosophic thought in
its most broad and general outlines, with which alone
we can deal.

And first let us consider, from this point of view, the
empirical or phenomenal theory of science described
in the last few pages. What is its metaphysical
import ? Does it impel us to philosophic agnosticism ?
Must we conclude that, if no scientific knowledge is
possible of any reality behind phenomena, no knowledge of any kind is there attainable ?

In reply, many philosophers would insist on the facts
that it is possible for us to construct a conceptual
model of nature consistent with itself, and consistent
to such an amazing extent with our sense-perceptions,
and hold that these facts are valid metaphysical
arguments that some reality exists outside our minds,
which conforms in some essential way to the picture
we frame of it. True that matter must be very
different from all our ideas of it, true the physical
nature of energy is incomprehensible, still matter and
energy are probably somethings between which exist
real relations corresponding to the relations postulated
by our scheme of physics, and confirmed in the world
of sense-perceptions. Thus the relations at all events

are realities. Electric current and electromotive force may be purely conceptual quantities, but the proportionality between them probably corresponds to some real property of that external world about which pure science may tell us nothing, but in which metaphysical insight insists on belief, and to which it calls science to witness in an alien court.

Many, too, would say that the fact that the human mind finds reason in nature is not purely because it is looking at itself projected there, not purely because, itself being reasonable, it sorts out and classifies only those relations in which it can find its own reason reflected. A human mind can conceive of chaos ; it can find chaos easily enough when it tackles a new problem, and often leaves it in chaos at the last. Since it finds reason and order in nature, then, it may fairly conclude that nature itself is orderly, that perhaps after all, in some faint way, natural law has points of likeness to legal ordinance, and may denote a lawgiver. This is not science, it is metaphysics once more calling on science to witness. But metaphysically the argument has weight ; those who do not wish to see all the wonders of creation concentrated in the human mind may still hold, in the old realistic no less than in the new conceptual sense, that " the heavens declare the glory of God, and the firmament showeth his handiwork."

During part of the nineteenth century, the idealistic philosophies of such men as Kant and Hegel stood in sharp contrast with the confident naturalism of those who, like Herbert Spencer, took the impress of the dominant school of scientific thought, and carried it over to metaphysics.

But, as the century drew to a close, the change in science was paralleled by a corresponding change in philosophy. The critical examination by Arthur Balfour of the basis of knowledge, the passing over of the majority of academic metaphysicians to the idealistic camp, the constructive pragmatism of William James with its idea of value as the criterion of truth, the " creative evolution " of Henri Bergson, are in one aspect but steps in the same direction, though not upon the same road.

Bergson's philosophy of change, though unacceptable to most metaphysicians, corresponds closely to the new vitalistic tendencies in biology. With his wealth of illustration drawn from all our knowledge of inorganic and organic nature, Bergson is far removed from the purely *a priori* philosophers of the older idealism, while his conclusions are even farther distant from those of the naturalistic philosophers who took the typical science of the mid nineteenth century as their source of inspiration.

The autonomy of life, the purposefulness of the organism, which, as we have seen, have so impressed themselves on some contemporary biologists, are to Bergson the corner-stone of his system. Behind such phenomena must lie a super-consciousness, free, indeterminate and incalculable. Breaking into matter, which is perhaps only a by-product of the creative impulse, as and where it can, it endows it with some share of its life—a share trammelled and enmeshed in matter, but still preserving some of the attributes of its free and unconfined source.

This process of becoming, this exaltation of the act of change, abolishes the timeless absolute of older

idealisms.   To Bergson time is of supreme significance.   In it are duration and change, the present
ever eating into the future, the past persisting into
the present.   Creative evolution is reality ; the vital
principle presses into new and ever new forms, failing
here and coming to a stop there, but ever pressing
onward, as far from teleology as from mechanical determinism, with " no goal," according to Mr Balfour,
"more definite than that of acquiring an ever fuller
volume of free creative energy."

Life uses matter as its vehicle ; calls on the stored
energy of plant life to support the vital output of the
animal ; fires the explosive material by a touch of
the trigger, and directs the forces thus set at liberty ;
or times the action of the mechanism to take
advantage of the chance accumulations of molecular
energy in the manner of Maxwell's dæmon.   As
Lodge has pointed out, such action is in full accord
with the physical principle of the conservation of
energy.   No work need be done in the processes of
timing and direction.

To Bergson, reason is chiefly concerned not with life
and freedom, but with the determinate mechanism
only introduced into the circle of life by its entangle-
ment with matter.   Thus it is that reason is most
at home with material and mechanical conceptions.
It has been evolved to enable us to deal with matter,
the waste product of creation.   Hence, in some ways,
instinct is nearer reality than reason.   Man touches
reality in those rare moments of crisis when emotion
and insight are fused with intuitive judgment, and his
whole being is alive with the will to act.   Then he
knows true freedom, " then he consciously sweeps

along with the advancing wave of Time, which as it moves creates."

Now, however fanciful this conception, however dangerous the exaltation of instinct over calm reason, it is impossible to overlook its analogy with the corresponding tendency in psychology. There, too, the reason of man has been found wanting as a complete explanation of his conduct, his character or his beliefs. Once more Bergson is in close touch with scientific tendencies, though many will hold that he carries them to an unwarranted extreme.

It is of interest to note that Bergson's philosophy is used by some theologians as an instrument of attack on rationalism, and by the revolutionary syndicalists of the Continent as a justification for instinctive attacks on the social order, when they have no reasoned account to give of what could take its place. So close in modern life are the cross-connections between different branches of human endeavour.

It is possible that danger to science as to society lies ahead. The dominance of the Universalist Roman Church nearly stifled the incipient science of the Northern race at the Renaissance; the dominance of the Universal proletariat, which some dread and others acclaim—a proletariat not dissimilar in race to the Southern rulers of the Roman Church—may threaten in the future the freedom of enquiry, the fearless exercise of reason, the full development of personality, that form the life-blood of the Northern race and its scientific achievement. The Roman Church saw its dogmas threatened by the new learning, and invoked torture and stake in an attempt to consolidate its forces. If the same race once more gains the ascendancy in Northern lands, as, by the differential birth-rate and the downward shift of political power, it seems destined to do, it is difficult to believe that scientific results which threaten its prejudices or are not in accord with its ideals will be respected.   Science, especially in its newly won field of sociology, might find itself once more in an environment where free and healthy growth was impossible.   Indications are not altogether wanting to remind us that reason is not always the mainspring of human action, and that the reign of ignorance and prejudice may yet again descend as a devastating blight on the human mind.

We have reached the end of our journey; we stand at the frontier, where the country opened up and partly surveyed by science touches the dark forest of the unknown future. What will the coming years disclose? No man can say.   He would be indeed bold who would hazard more than a guess at the direction of the next few steps.

Science and the Human Mind.

Yet the story of the development of natural knowledge has an importance and a significance beyond its mere historic interest, great though that may be. The story of how discoveries were made in the past throws light on the means by which further advance becomes possible, and gives a clearer idea of the inner meaning of what is already known than can be obtained by non-historical methods.

Scientific laws and theories, when taken from the conceptual sphere in which they may be rigidly valid, and referred back to the interpretation of nature which suggested them, are essentially provisional. They are but working hypotheses adapted to the needs of the passing age. What part of the conceptual model seemed better adapted to stand as a permanent representation of nature than the immutability of the chemical elements, what surer than the Newtonian dynamics ? Yet, under the pressure of radio-active phenomena, we have been forced to repudiate the one as a universal law, and to modify the other to explain the motion of bodies which approach too nearly the velocity of light.

Again and again in the course of the preceding chapters we have traced how a theory, put forward at a definite stage in the development of a subject, interprets adequately and successfully existing knowledge and points the way for future research. Yet, a few pages later, we are driven to recount its failure under the test of ever-advancing thought, and its supersession by some other hypothesis more adapted to the needs of the new time, and more fruitful in suggestions for further work.

Moreover, we have had continually to acknowledge that the later theory would have been less useful at the previous stage than the outworn idea it replaced. We come to see that the test of a successful theory is not its concordance with absolute truth—with which indeed we can never compare it—but the humbler and more practically useful rôle of giving us a means of co-ordinating conveniently and succinctly our existing unconnected pieces of knowledge, and of enabling us to frame experiments and enquiries calculated to open up the greatest extent of hitherto unsurveyed territory in the kingdom of nature. The test of a theory, then, is whether or no in its day it is a satisfactory " working hypothesis."

The history of the caloric theory of heat is a good instance of the successive changes in value of a scientific idea. This theory of a weightless fluid gave a vivid conception of heat as a quantity to be measured, and was indispensable to the earlier workers in calorimetry. It failed to explain the unlimited development of heat by friction, and was accordingly laid on one side during the investigation of the mechanical aspect of heat studied in the form of energy. But now that further research has analysed the concept of energy into the two factors of intensity and quantity, there are signs that, as applied to the latter, a modification of the old caloric theory may once more find a place in the modern structure.

It is needless to repeat instances. Scientific theories certainly " mount on stepping stones of their dead selves to higher things." At our present stage, a common basis of matter, energy as a quantity constant in amount, evolution by means of natural selection, are working hypotheses co-ordinating much of our knowledge and underlying almost all our researches into the unknown. They are suited to the age and to the minds of present investigators, and we have not yet exhausted their possibilities. But, in the light of history, it would be rash to proclaim them as absolute, eternal truth. We may hereafter discover new kinds of corpuscles and again for a time have to give up the idea of one basis of matter. Energy may prove to be conserved only in the limited conditions hitherto studied, and natural selection fail to explain all the wonders of creative evolution. Such a thought need not hinder the fullest use of those conceptions in our present stage ; for us, and now, they are true—true enough, anyhow. But we must keep an open mind in the future, and beware of blocking an advance by a blind reliance on the authority of theories which, by such treatment, are elevated or degraded into dogmas.

So too with those wider philosophical questions that for ever rise from scientific problems. The successive oscillations of thought from mechanical to vitalistic tendencies has been one of the features of our survey. At recurrent times, the best work is done under the inspiration of the hope of unifying the whole of nature in one comprehensive scheme. The Greek atomists extended their conceptions of the inorganic world to cover the phenomena of life, unconscious of the logical chasms in their reasoning. Some rash materialists thought that natural selection had solved all problems. Once and again the explanation proved insufficient, and the hope of reducing everything to one science of nature, for the time at any rate, has had to be abandoned. We have to be content with regarding nature from several separate aspects, and console ourselves for the loss of unity in the thought of the greater fullness and richness of the manifold prospect.

Considered mechanically, a man, for instance, is a somewhat complicated piece of mechanism of certain dimensions, containing levers of various kinds and sizes, and conforming to all the conceptual laws of mechanics, such as gravity, like any inorganic body. To the chemist he is a chemical laboratory, in which many changes go on in various organs, to be represented ultimately by combinations and recombinations among the atoms and molecules of the different chemical elements which make up his frame.   To the physicist he illustrates processes such as osmosis or electrolytic conduction, and the atoms of the chemist are resolved into corpuscles or electrons, the vibrations of which emit the electromagnetic radiation known as radiant heat.   By the physiologist he is resolved into a collection of cells, and the changes studied by the physicist and chemist are considered in their bearing on the general life of the organism.   To the anthropologist and zoologist the man is an individual of a certain race which can be placed in its due class among the other races of men and amid the long sequence of geological specimens of other animals.   To the psychologist the man is primarily a mind, and his typical product is a thought.   To his doctor he is an obscure and ill-understood piece of machinery, mechanical, chemical and psychological ; while to his vicar he is essentially an immortal soul to be saved or lost.

Each action of the man may be dealt with from many points of view.   One of his thoughts to the parson may be a sin, to the psychologist an illustration of the effect of suggestion, to the physiologist a function of the grey matter of his brain, while to the physicist and chemist the thought may be represented by electrical or chemical changes in those cerebral cells.

In the present state of science, it is impossible to say that any one of these aspects is more fundamental than the others, even if we reduce them to the three main aspects of physics, biology and psychology.

It is impossible to say which *is* the man or the thought. The different aspects are on different planes, incommensurable. A thought is represented by physical and physiological changes, but it is none the less a psychological phenomenon, part of the consciousness of the man and part of his spiritual life.

If, in the distant future, science should approach towards a complete explanation of a man regarded from without, expressed in electrical, mechanical or æthereal terms, it would be within sight of unity in the scientific domain—of that unity which has been the dream of so many philosophers and men of science. Yet, inwardly, the consciousness of that man would to himself remain an unconquered citadel, a kingdom of heaven within him.

Without necessarily holding the view of Kant and his modern followers that the Ego prescribes its own laws to nature, and sees its own reason reflected in natural phenomena, it is clear that the particular scheme of theories in which modern science is expressed must be regarded as conditional on, and indeed originated by, the constitution and peculiarities of our minds. As we have said when dealing with the history of these conceptions, the feeling that an explanation is more satisfying and fundamental when expressed in mechanical terms is probably a consequence of the fact that our minds and bodies possess a definite muscular sense, so that for us force is a direct sense-perception. Did we, like the fish-torpedo, possess a special electric sense, we might regard electric conceptions as equally or more fundamental and satisfying than mechanics. Moreover, modern science in its essence has been formulated and developed by the races of North-Western Europe. The scientific conceptions suitable to interpret natural (and indeed supernatural) phenomena to our race are not necessarily suitable for other races. Modern science might have worn a different guise had it developed in other peoples, perhaps would never have developed at all had not the Northern races been impelled to investigate nature experimentally at the right stage in the history of the world.

But when all criticism is done, we cannot but pause, wondering and amazed at the majesty of the temple of science. Whether we stand within and gaze at the beauty of its proportions and their concord one with another, or whether we pass without and trace the success with which it fits and interprets the spirit of the landscape, we are equally fain to confess that it is the grandest work of the human intellect. And, like a cathedral that has been a shrine for many generations of the sons of men, it can be altered to changing needs. No rearrangement of parts of the superstructure, no addition of mighty tower or soaring spire, need endanger its foundations. Firm on the rock of experience, free from the shifting sands of metaphysical systems, with room on all sides for new aisles and chapels and altars, it stands, a triumph of truth and patient perseverance, and an eternal sanctuary for the human mind.

BIBLIOGRAPHY

It is impossible to give a complete account of the books and papers to which the authors are indebted in the preparation of this work. The following list is prepared primarily as an indication where more detailed information on special points may be found. Those works suitable for the general reader are marked with an asterisk.

CHAPTER I

INTRODUCTION

Also for general reference throughout the book.

History of the Inductive Sciences. W. Whewell. 3 vols., 1857.

History of Civilization in England. H. T. Buckle. 3rd ed., 3 vols., 1871.

History of Philosophy. G. H. Lewes. 5th ed., 2 vols., 1880.

History of Materialism. F. A. Lange. Eng. trans., 3 vols., 1877.

*The Foundations of the Nineteenth Century. H. S. Chamberlain. Eng. trans., 2 vols., 1910.

History of the Philosophy of History. Robert Flint. 1893.

*The Races of Europe. W. L. Ripley. 1900.

*History of Mathematics. W. W. Rouse Ball. London, 1901.

Die Mechanik in ihre Entwickelung. E. Mach. 1883. *Eng. trans., 2nd ed., 1902.

*History of Astronomy. A. Berry. 1898.

Geschichte der Physik. J. C. Poggendorff. 1879.

*A History of Chemical Theories and Laws. M. M. Pattison Muir. 1907.
The Study of Chemical Composition : Its Method and Historical Development. Ida Freund. 1904.
Klassiker der exacten Wissenschaften. W. Ostwald. 1889, etc.
*Medical History from the Earliest Times. E. T. Withington. 1894.
*The Science of Life : an Outline of the History of Biology. J. A. Thomson. (c. 1900.)
*The Cambridge Modern History, the Dictionary of National Biography, and the *Encyclopædia Britannica, 10th and 11th editions, have been used freely throughout the work. References are only given for some special reason

CHAPTER II

SCIENCE IN THE ANCIENT WORLD

*The Dawn of Civilization. G. Maspero. Eng. trans., 1896.
Religions of Ancient Egypt and Babylonia. A. H. Sayce. 1902.
Die Ägyptologie. H. Brugsch. 1891.
The Religion of Ancient Egypt. W. M. Flinders Petrie. 1906.
*Ægean Civilisation. Crete. Encyclopædia Britannica, 11th ed.
A History of Ancient Greek Literature. Gilbert Murray. 1897.
Early Age of Greece. W. Ridgway. 1901.
*The Greek View of Life. G. Lowes Dickinson. 1896.
History of Greek Philosophy. Ed. Zeller. Eng. trans., 1881.
*Greek Thinkers. T. Gomperz. Eng. trans., 3 vols., 1901–5.
Histoire de la Philosophie Atomistique. Mabilleau. 1895.

The Atomic Theory of Lucretius. J. Masson. 1884.

De Rerum Natura. Lucretius. Ed. H. A. J. Munro. 2 vols., 1866.

The Works of Aristotle, translated into English. Clarendon Press (now appearing).

The Works of Archimedes. T. L. Heath. 1897.

*Medicine. T. Clifford Allbutt. Encyclopædia Britannica, 11th ed.

The Decline and Fall of the Roman Empire. E. Gibbon. Ed. J. B. Bury. 7 vols., 1896.

La Cité Antique. F. de Coulanges. 1898.

The Greatness and Decline of Rome. G. Ferrero. Eng. trans., 2 vols., 1907.

Letters of the Younger Pliny. Ed. J. D. Lewis. 1879.

*Astronomy. Encyclopædia Britannica.

Foundations of the Nineteenth Century. H. S. Chamberlain. 1910. Ch. V.: The Entry of the Jews.

History of the Jews. H. H. Milman. 1830.

CHAPTER III

THE MEDIÆVAL MIND

Cambridge Mediæval History. I. Ch. XX.: Thought and Ideas. H. F. Stewart.

*The Mediæval Mind. H. O. Taylor. 2 vols., 1911.

La Connaissance de la Nature et du Monde au Moyen Age. Ch. V. Langlois. 1911.

*The Universities of Europe in the Middle Ages. H. Rashdall. 3 vols., 1895.

Illustrations of the History of Mediæval Thought. R. L. Poole. 1884.

A Mediæval Garner. G. G. Coulton. 1911.

History of Christianity under the Empire. H. H. Milman. 1840.

History of Latin Christianity. H. H. Milman. 1855.

19

Histoire des Origines du Christianisme. E. Renan. 8 vols., 1866–83.

*The Growth of Christianity. Percy Gardner. 1907.

The Making of Western Europe. Vol. I.—The Dark Ages. C. R. L. Fletcher. 1912.

The Dark Ages (European Literature). W. P. Ker. 1894.

The Earthly Paradise. William Morris. 1868–1870.

Sigurd the Volsung. William Morris. 1876.

The Story of Burnt Njal. Sir George Dasent. 1900.

Mohammedanism. Article in Encyclopædia Britannica.

Malaria and Greek History. W. H. S. Jones. 1909.

The Greatness and Decline of Rome. G. Ferrero. Eng. trans., 2 vols., 1907.

Bede's Ecclesiastical History of England and the Anglo-Saxon Chronicle. Bohn's Antiquarian Library. 1871.

*The Goths. H. Bradley. 1891.

*Italy and her Invaders. T. Hodgkin. 2nd ed., 1892.

L'Islamism et la Science. E. Renan. 1883.

Averroès et l'Averroïsme. E. Renan. 2nd ed., 1861.

Thomas Aquinas. Opera. Parma. 25 vols., 1852–73.

Thomas Aquinas. Of God and His Creatures. Abridged trans. of Summa contra Gentiles. J. J. Rickaby, S.J. 1905.

New Things and Old in Saint Thomas Aquinas. H. C. O'Neill. 1909.

*Robert Grosseteste. F. S. Stevenson. 1899.

Roger Bacon. Opera. (Rolls Series.) Ed. J. S. Brewer. 1859.

Roger Bacon. Sa Vie, ses Ouvrages, ses Doctrines. E. Charles. 1861.

Roger Bacon. Eine Monographie. Schneider. 1873.

Roger Bacon. Opus Magnus. Ed. by J. H. Bridges. 1897.

Martin Luther and the Reformation in Germany. Charles Beard. 1889.

CHAPTER IV

THE RENAISSANCE AND ITS ACHIEVEMENT

*Cambridge Modern History (1902–1911). Vol. I.
  Ch. XVI.: The Classical Renaissance. R. C. Jebb.
  Ch. XVII.: The Christian Renaissance. M. R. James.
*The Renaissance in Italy. J. A. Symonds. 1897.
Life of Leo X. W. Roscoe. 1805.
*Léonard de Vinci. Gabriel Séailles. 1906.
Leonardo da Vinci's Note Books. Arranged and rendered
  into English by Ed. M'Curdy. 1906.
La Philosophie de Léonard de Vinci. Peladan. 1910.
Giordano Bruno. J. L. M'Intyre. 1903.
Life of Giordano Bruno. T. Firth. 1887.
The Praise of Folly. Erasmus. New ed., 1870.
History of the Conflict between Religion and Science.
  J. W. Draper. 1875.
*Galileo, his Life and Work. J. J. Fahie. 1903.
De Magnete. W. Gilbert. 1600. Eng. trans., ed. S. P.
  Thompson. 1900.
Physiologia Epicuro-Gassendi-Charltoniana; or, A Fabric
  of Science Natural. Walter Charlton. 1654.
Bacon, Francis, Viscount St Albans. Letters and Life.
  J. Spedding. 7 vols., 1861–74.
*Bacon's Essays and the Advancement of Learning. 1902.
Francis Bacon. Works. Ed. Ellis, Spedding and Heath.
  1857–70.
Essaies de Montaigne. New ed., 7 vols., 1886–89.
Cartesianism. Encyclopædia Britannica.
Port Royal. C. A. Sainte-Beuve. 6 vols., 1882.
Port Royal. Charles Beard. 2 vols., 1861.
Lettres Provinciales. Blaise Pascal. New ed., 1853.
Types of Ethical Theory. James Martineau. 2 vols.,
  1886.
*Sir Isaac Newton. Life. D. Brewster. 2 vols., 1855.
  Opera. Ed. S. Horsley. 5 vols., 1779–85.

Essay on Newton's Principia. W. W. Rouse Ball. 1893.

*Cambridge Modern History. Vol. V., Ch. XXIII.: European Science. W. W. Rouse Ball and Sir Michael Foster. 1908.

The Sceptical Chemist. Robert Boyle. 1680.

Boyle, Hon. Robert. Works. Thomas Birch. 5 vols., 1744.

*Principal Navigations, Voyages, Traffiques and Discoveries. R. Hakluyt. 3 vols., 1598-1600.

New Voyage round the World. By Captain Wm. Dampier. 4 vols., 1703.

Herbal. J. Gerard. 1597.

*History of the Rise and Influence of the Spirit of Rationalism in Europe. W. E. H. Lecky. 2nd ed., 2 vols., 1865.

Christian Mysticism. W. R. Inge. 1899.

Select Discourses of John Smith. H. G. Williams. 1859.

The True Intellectual System of the Universe. Ralph Cudworth. 3 vols., 1845.

Cambridge Modern History. Vol. VI., Ch. XXIII.: English Political Philosophy in the Seventeenth and Eighteenth Centuries. A. L. Smith. 1909.

CHAPTER V

THE PHYSICS OF THE NINETEENTH CENTURY

Lectures on the French Revolution. By Lord Acton. Ed. J. N. Figgis and R. V. Lawrence. 1910.

Diderot and the Encyclopædists. John Morley. 1878.

A History of Modern Philosophy from the Close of the Renaissance to 1880. H. Höffding. Eng. trans., 1900.

A History of European Thought in the Nineteenth Century. J. T. Merz. 2 vols., 1896.

Mechanics. J. Cox. 1904.

*History of Astronomy during the Nineteenth Century. Agnes M. Clerke. 4th ed., 1903.

The Electrical Researches of the Hon. Henry Cavendish. Ed. by J. Clerk Maxwell. 1878.

The Theory of Experimental Electricity. W. C. D. Whetham. 2nd ed., 1912.

The Principles of Chemistry. D. Mendeléeff. 1891.

Theory of Heat. J. Clerk Maxwell. 1880.

Theory of Light. T. Preston. 3rd ed., 1901.

Theory of Heat. T. Preston. 2nd ed., 1904.

Heat. J. H. Poynting and J. J. Thomson. 4th ed., 1911.

Thermodynamics. J. Parker. 1894.

Outline of the Theory of Thermodynamics. E. Buckingham. 1900.

*Experimental Researches in Electricity. Michael Faraday. 3 vols., 1844-55.

Bakerian Lectures. Sir Humphry Davy. Philosophical Transactions of the Royal Society of London. 1806, 1808 and 1809.

Electricity and Magnetism. J. Clerk Maxwell. 1873.

A History of the Theories of Æther and Electricity. E. T. Whittaker. 1910.

Theory of Solution. W. C. D. Whetham. 1902.

CHAPTER VI

THE COMING OF EVOLUTION

*Darwin and Modern Science. Ed. A. C. Seward. 1909.

From the Greeks to Darwin. H. F. Osborn. 1894.

*The Coming of Evolution. J. W. Judd. 1910.

The Voyage of the "Beagle." Natural History, Geology, etc. 1839-45.

Reports of Committees of the Royal Society of London.

*Principles of Geology. Sir Charles Lyell. 12th ed., 2 vols., 1875.

Synthetic Philosophy. Herbert Spencer. 9 vols., 1860–93.
Principles of Population. T. R. Malthus. 1798.
*The Origin of Species. Charles Darwin. 2 vols., 6th
ed., 1891.
*The Descent of Man. Charles Darwin. 1871.
*Life and Letters of Charles Darwin. Ed. by Francis
Darwin. 3 vols., 1887.
*More Letters of Charles Darwin. Francis Darwin and
A. C. Seward. 1903.
*Life and Letters of Thomas Henry Huxley. Leonard
Huxley. 1900.
*Man's Place in Nature. T. H. Huxley. 1863.
Origin of Civilisation. Sir John Lubbock. 1870. 6th
ed., 1902.
*Primitive Culture. E. B. Tylor. 4th ed., 2 vols., 1903.
Anthropology. E. B. Tylor. 1881.
*Hereditary Genius. Francis Galton. 1869.
Human Faculty. Francis Galton. 1883.
Natural Inheritance. Francis Galton. 1889.
The Laws of Heredity. G. Archdall Reid. 1911.
*Introduction to Eugenics. W. C. D. and C. D. Whetham.
1912.
*The Family and the Nation. W. C. D. and C. D. Whetham.
1909.
Thus spake Zarathustra. F. W. Nietzsche. Eng. trans.,
1902.
F. W. Nietzsche. Works. Eng. trans., 1896, etc.
The Riddle of the Universe. E. Haeckel. 1900.
Social Evolution. Benjamin Kidd. 1894.
Evidences of Christianity. W. Paley. 1794.

CHAPTER VII

THE LATEST STAGE

*The Recent Development of Physical Science. W. C. D.
Whetham. 4th ed., 1910.

The Conduction of Electricity through Gases. J. J. Thomson. 2nd ed., 1906.

Radioactivity. E. Rutherford. 2nd ed., 1905.

Æther and Matter. J. Larmor. 1900.

Physical Optics. R. W. Wood. 1905.

Modern Electrical Theory. N. R. Campbell. 1907.

Zur Elektrodynamik bewegter Körper. Annalen der Physik, xvii. A. Einstein. 1905.

The Common Sense of Relativity. Philosophical Magazine, xxi. N. R. Campbell. 1911.

Mendel's Principles of Heredity. W. Bateson. 1909.

*Mendelism. R. C. Punnett. 3rd ed., 1911.

*Heredity and Society. W. C. D. and C. D. Whetham. 1912.

The Science and Philosophy of the Organism. H. Driesch. 1907.

*Is there One Science of Nature ? Hibbert Journal, Oct. 1911 and Jan. 1912. J. A. Thomson.

*The Theory of Life. Presidential Address to the British Association. E. A. Schäfer. 1912.

*Experimental Psychology. C. S. Myers. 1911.

A Manual of Psychology. G. F. Stout. 2nd ed., 1901.

Psychology. J. Ward. Encyclopædia Britannica.

*Psychical Research. W. R. Barrett. 1912.

Proceedings of the Society of Psychical Research. 1882, etc.

*Human Nature in Politics. Graham Wallas. 2nd ed., 1910.

*The Problems of Philosophy. Hon. Bertrand Russell. 1911.

*Introduction to Mathematics. A. N. Whitehead. 1911.

A System of Logic. J. S. Mill. 9th ed., 2 vols., 1875.

On Liberty. J. S. Mill. 1859.

On Representative Government. J. S. Mill. 3rd ed., 1865.

The English Utilitarians. Leslie Stephen. 3 vols., 1900.

*Anthropology. R. R. Marett. 1911.

*The Golden Bough. J. G. Frazer. 2nd ed., 3 vols., 1900.

Themis. Jane Harrison. 1911.

From Religion to Philosophy. F. M. Cornford. 1912.

Philosophy as Scientia Scientiarum. Robert Flint. 1904.

*The Grammar of Science. Karl Pearson. 2nd ed., 1900.

La Science et l'Hypothèse. Henri Poincaré. 1902.

La Valeur de la Science. Henri Poincaré. 1905.

Scientific Fact and Metaphysical Reality. R. B. Arnold. 1904.

*The Foundations of Science. W. C. D. Whetham. 1912.

Berkeley's Works. Ed. A. C. Frazer. 4 vols., 1901.

Naturalism and Agnosticism. James Ward. 2nd. ed., 2 vols., 1903.

A Study of Religion. James Martineau. 2 vols., 1889.

A Defence of Philosophic Doubt. A. J. Balfour. 1879.

The Foundations of Belief. A. J. Balfour. 3rd ed., 1895.

The Will to Believe. William James. 1897.

*Varieties of Religious Experience. William James. 1908.

Creative Evolution. Henri Bergson. Eng. trans., 1911.

Matter and Memory. Henri Bergson. Eng. trans., 1911.

Philosophy of Bergson. A. D. Lindsay. 1911.

*Life and Consciousness. Hibbert Journal, October 1911. H. Bergson.

*Creative Evolution and Philosophic Doubt. Hibbert Journal, Oct. 1911. A. J. Balfour.

Treatise on Thermodynamics. M. Planck. Eng. trans., 1903.

The Caloric Theory of Heat. Address to Section A, Brit. Association. H. L. Callendar. 1912.

INDEX

PRINTED BY NEILL AND CO., LTD., EDINBURGH

ABOUT THE AUTHOR

SIR WILLIAM CECIL DAMPIER

(*Formerly,* WHETHAM)

(February, 1818 – February 20, 1895)

Sir William Cecil Dampier FRS (born William Cecil Dampier Whetham) (December 27, 1867 – December 11, 1952) was a British scientist, agriculturist, and science historian who developed a method of extracting lactose (milk sugar) from whey.

He was born in London, the son of Charles Langley and Mary (née Dampier) Whetham and the grandson of Sir Charles Whetham, a former Lord Mayor of London. In 1886, he entered Trinity College, Cambridge and in 1889 commenced his varied researches in the Cavendish Laboratory. In 1891 was elected a Fellow of Trinity.

In June 1901 he was elected a Fellow of the Royal Society. His candidacy citation read: *"Lecturer in Physics. Fellow of Trinity College, Cambridge.*

Dampier was author of the following scientific papers, &c: - 'On the Alleged Slipping at the Boundary of a Liquid in Motion'; 'Note on Kohlrausch's Theory of Ionic Velocity'; 'Ionic Velocities'; 'On the velocity of the Hydrogen Ion through Solutions of Acetates'; 'On the Velocities of the Ions and the Relative Ionization Powers of Solvents';[6] 'The Velocity of the Ions'; 'The Ionizing Power of Solvents';[8] 'Report to the British Association on the Present State of our Knowledge in Electrolysis and Electro-Chemistry'; 'The Theory of the Migration of the ions and of Specific Ionic Velocities'; 'The Coagulative Power of Electrolytes'; 'The Ionization of Dilute Solutions at the Freezing Point' (a paper read before the Royal Society); an elementary text book on 'Solution and Electrolysis'; Letters and Articles in 'Nature' and 'Science Progress.'"

In 1904 he published the first of his broader works on science and its history, *The Recent Development of Physical Science*. This was followed in 1929 by his frequently reprinted and translated *A History of Science, and its Relations with Philosophy and Religion*. *A Shorter History of Science*. 1944,1945.

From 1931 to 1935 he served as the first secretary of the Agricultural Research Council. He was knighted in 1931 for public service to agriculture.

Family

In 1897 he had married Catherine Durning Holt of a Liverpool shipowning family. They had one son, and four daughters, including Margaret Dampier Whetham, Elizabeth Cockburn and Edith Holt Whetham.

FEEDBACK

Now that you have read the book ...

Was it interesting?

Did you enjoy what you wanted to read?
Was there any room for improvement?

Let us know at:
http://www.diamondbooks.ca/feedback

Your feedback is highly appreciated.
Thank you!

Would you like to buy a copy of 'GREAT ASTRONOMERS' ?

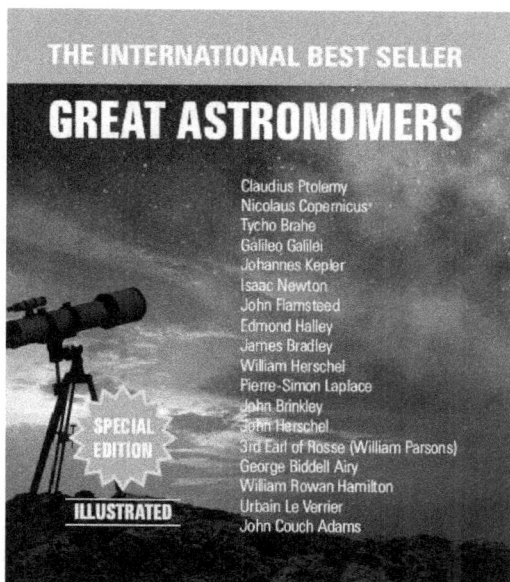

THE INTERNATIONAL BEST SELLER

GREAT ASTRONOMERS

Claudius Ptolemy
Nicolaus Copernicus
Tycho Brahe
Galileo Galilei
Johannes Kepler
Isaac Newton
John Flamsteed
Edmond Halley
James Bradley
William Herschel
Pierre-Simon Laplace
John Brinkley
John Herschel
3rd Earl of Rosse (William Parsons)
George Biddell Airy
William Rowan Hamilton
Urbain Le Verrier
John Couch Adams

SPECIAL EDITION

ILLUSTRATED

Robert Stawell Ball (July 1, 1840 – November 25, 1913)
AN IRISH ASTRONOMER WHO FOUNDED THE SCREW THEORY.

ROBERT STAWELL BALL

Original Edition

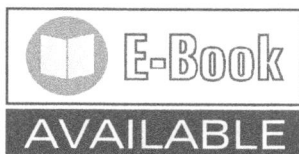

E-Book
AVAILABLE

Please visit:
http://www.diamondbooks.ca/books

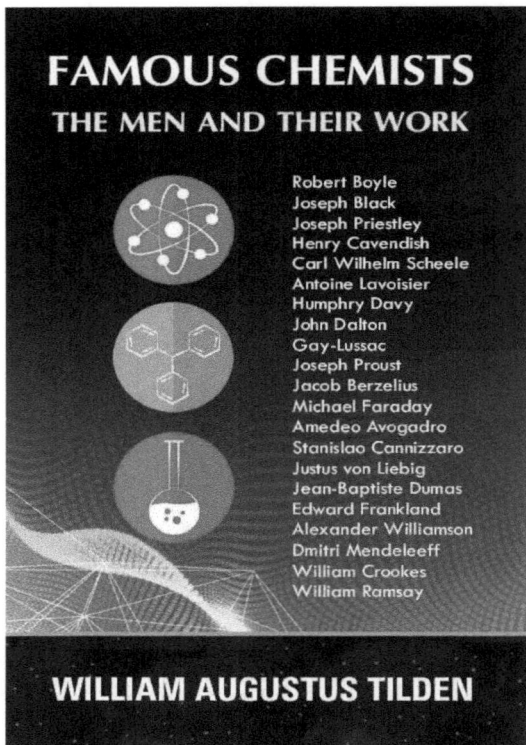

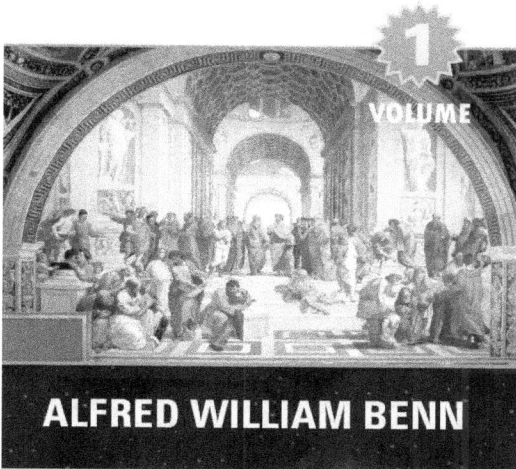

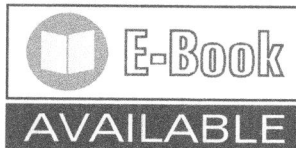

Would you like to buy a copy of
'THE GREEK PHILOSOPHERS' ?

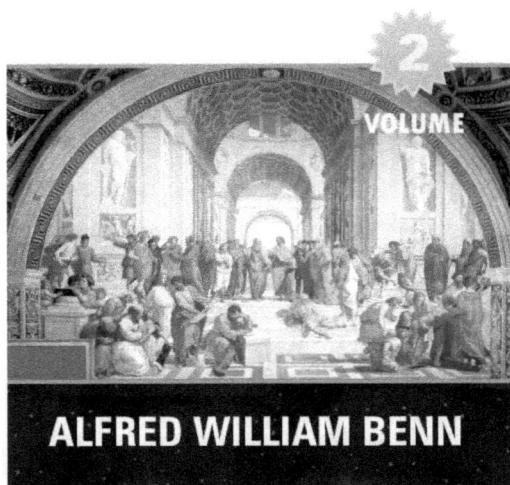

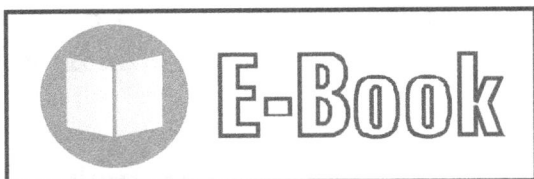

E-Book

AVAILABLE
FOR ALL PUBLISHED TITLES

Please visit:
http://www.diamondbooks.ca/books

DIAMOND
BOOKS
www.diamondbooks.ca

HUGE SAVINGS ON BULK ORDERS
(10 copies, 20 copies, 50 copies, 100 copies, 500 copies, 1000 copies)

Please send your request at:
http://www.diamondbooks.ca/bulkorder